APOLLO
11

APOLLO 11

THE INSIDE STORY

DAVID WHITEHOUSE

ICON

Published in the UK in 2019
by Icon Books Ltd, Omnibus Business Centre,
39–41 North Road, London N7 9DP
email: info@iconbooks.com
www.iconbooks.com

Sold in the UK, Europe and Asia
by Faber & Faber Ltd, Bloomsbury House,
74–77 Great Russell Street,
London WC1B 3DA or their agents

Distributed in the UK, Europe and Asia
by Grantham Book Services,
Trent Road, Grantham NG31 7XQ

Distributed in the USA
by Publishers Group West,
1700 Fourth Street, Berkeley, CA 94710

Distributed in Australia and New Zealand
by Allen & Unwin Pty Ltd,
PO Box 8500, 83 Alexander Street,
Crows Nest, NSW 2065

Distributed in South Africa
by Jonathan Ball, Office B4, The District,
41 Sir Lowry Road, Woodstock 7925

Distributed in India by Penguin Books India,
7th Floor, Infinity Tower – C, DLF Cyber City,
Gurgaon 122002, Haryana

Distributed in Canada by Publishers Group Canada,
76 Stafford Street, Unit 300
Toronto, Ontario M6J 2S1

ISBN 978-178578-512-2

Typeset in Electra by Marie Doherty

Printed and bound in Great Britain
by Clays Ltd, Elcograf S.p.A.

To Jill, as well as the Moon and the stars

Contents

Acknowledgements

It was my agent Laura Susijn who suggested I write this book. I, like many others, knew the 50th anniversary of the first Moon landing was approaching, and I anticipated many books about it would be written. I was a little reluctant to be one among many. But then I looked back into my archive, by which I mean sealed cardboard boxes stored in my loft. I soon realized that it was a goldmine of information. Over the more than 40 years since I was a young man I had been collecting. As I started my career as an astronomer at the Jodrell Bank radio observatory, I began to meet astronauts, engineers and officials involved in the Apollo project. Such meetings increased as I moved to the University of London's Space Science department. As I became involved in the media I started to be invited to receptions, press conferences and dinners with a growing number of my childhood heroes. In 1988 I joined the BBC as science correspondent and soon realized it was a job that opened doors, and that people took my calls.

Some astronauts, like Neil Armstrong, treated writers with suspicion. He disliked articles that featured him as a personality. Other astronauts, well, they could talk and talk. Often I would hear them tell the familiar story they had been giving to journalists for years; I sat through that and hoped they would open up when I showed I had a deeper knowledge than most other science journalists. I remember Alan Shepard did that, pausing with a mischievous glint in his eye when I asked him an unexpected question. Sometimes it didn't work. More than one cosmonaut who had been involved in their Moon program pulled down the shutters when I asked something they considered awkward. Over the years I met all the moonwalkers and interviewed most of them, along with a great number of other astronauts and cosmonauts, administrators and officials. My boxes were full of tapes, notepads, press kits and much other stuff. Combined with what is available in NASA's extensive archive, I decided there was a book to be written that put the people first and used, as far as possible, the words of those involved.

I would like to thank Neil Armstrong, Buzz Aldrin, Gene Cernan, David Scott, John Young, Alan Shepard, James Lovell, Charlie Duke, Donn Eisele, Alan Bean, Gordon Cooper, Al Worden, Walt Cunningham, Tom Stafford, Dick Gordon, John Glenn, Pete Conrad, Edgar Mitchell, Richard Gordon, James Irwin, Stu Roosa, Ron Evans, Deke Slayton, Wally Schirra, James Fletcher, Thomas Paine, Joe Shea, Rocco Petrone, Brainerd Holmes, Bob Gilruth, George Mueller, James Webb, John Houbolt, Robert Seamans, Max Faget, William Pickering, Sergei Khrushchev, Viktor Savinykh, Georgi Grechko, Yuri Romanenko and Pavel Popovich.

I thank Laura Susijn for believing in this book and all those at Icon Books who made it a reality; also Nick Booth, who is a constant source of advice on space matters and good writing. He knows the life of a writer. More thanks than I can ever express go to my wife Jill and my children, Christopher, Lucy and Emily.

The night of the first landing on the Moon my father initially said I had to go to bed as the first footprint was scheduled for 3 am. I eventually talked him around, and watched transfixed on our black-and-white TV. I have never gotten over that night. My parents are no longer with me, but I hope they realized what that meant to a young boy. I think they did. I am sad that my children have not seen the like.

'The Moon is a friend for the lonesome to talk to'
—NEIL ARMSTRONG

'We must master the highest technology or be crushed'
—VLADIMIR LENIN

Prologue

Spaceflight is dangerous. Everyone knew that. None more so than the astronauts, their families, and all those intimately involved in Apollo – the project to land a man on the Moon. On that evening, 20 July 1969, everyone in Mission Control in Houston knew the danger. The lives of the two men about to attempt a descent to the lunar surface depended upon single moments, the single decision any of them might have to make in a second.

Gene Kranz, 31 years old, was the Flight Director for the landing. He was confident in his abilities, though not arrogant. His job was to run the show by being able to assimilate all the information coming into Mission Control from 'Eagle', the Lunar Lander. Formerly a fighter pilot and an engineer, he was leading a talented group. In front of him were rows of computer screens at which sat the mission controllers, all with their individual roles and names such as Guido and EECOM. Kranz's call sign was Flight. The average age of the men of Mission Control was only 26.

He selected a private loop only heard by the controllers. He didn't want anyone else to listen to what he was about to say. He waited a second and spoke:

Today is our day, and the hopes and dreams of the entire world are with us. This is our time and our place, and we

will remember this day and what we do here always. In
the next hour we will do something that has never been
done before – we will land an American on the Moon.

Less than an hour after those words were spoken, Neil
Armstrong and Buzz Aldrin were 500 metres from the lunar
surface in the region of the Sabine complex of small craters on
the western shore of the Sea of Tranquillity. Armstrong was 38
and regarded as the best person to attempt the first landing. He
had to put Eagle down within the next three minutes. Next to
him was Edwin 'Buzz' Aldrin, who was 39 years old. Ahead of
them stretched the dark expanse of lunar night. No one, especially Armstrong himself, knew if they were going to make it.
The landing was the goal; the moonwalk was secondary. Before
he left Earth, he had told close colleagues that he had only a
50:50 chance of pulling off a successful landing.

In a few seconds Armstrong would have to take over and fly
Eagle manually while looking for a suitable site. It was the most
difficult thing any pilot had ever been asked to do. His heart
rate was 160, double normal. He flexed his right hand around
the joystick that controlled Eagle's attitude, and his left hand
around the thrust controller. He knew that this was never a job
for a computer. The windows on the Eagle overhung the floor
and were angled downwards so that he could lean against the
tethers that tied him to the floor and look downwards.

On the right side of his body was Eagle's control panel.
Dials, gauges, switches and lights concerning all aspects of the
Eagle from fuel to radar, attitude to rate of descent. Two but-
tons stood out, the only ones with a striped surround. One was

labelled 'Abort', which sent the Eagle back up into orbit, and the other 'Abort Stage', which initiated an even more danger-ous manoeuvre.

They were coming in from the east across the Sea of Tranquillity – a misnomer as there were only dry rocks beneath them – and with the low Sun behind them, the shadows were long and deeply dark. They had passed over rough ground a few minutes ago and were heading towards Tranquillity's south-western region. They were 'long', or downrange. Between the lumpy gravity of the Moon and some extra speed picked up when they undocked from the Apollo Command Module, Eagle was at least a second ahead of its timeline and that trans-lated into a mile too far. Armstrong had noticed the discrepancy immediately, Aldrin later said he hadn't and was impressed by his crewmate's alertness. But just one second could make all the difference. The rocks below looked terrifying and the computer was taking Eagle directly into them. Armstrong took over manual control at 150 m, not 50 m as had been planned. He needed more time to look for a smooth landing site.

There was only enough fuel for one landing attempt, and that was running out fast. They were flying for their lives. Aldrin was too busy to look out of the window – that was planned. He later said that if it wasn't on the dials he wasn't looking at it. He was looking upwards and to his left at the radar display with its altitude and rate of descent readings: 'Three hundred and fifty feet; down at four. Three hundred thirty, six and a half down. You're pegged on horizontal velocity.'

Back in Houston Mission Control, Kranz, in the centre console, knew they were close to an abort. He looked to his left where the Capcom – capsule communicator – Charlie Duke was sitting. He was the only one who spoke to the astronauts. To Duke's left was Jim Lovell and to Lovell's left was Fred Haise. Lovell had been the backup commander and as such if Armstrong were to have been injured just before the mission it would have been him doing the flying now. Fred Haise was Aldrin's backup. They had the procedures, the checklists and many documents before them but most of all they had simulator experience of the landing. Lowell thought Duke was talking too much so he tapped him on the shoulder suggesting fewer words: 'Let him do his job,' he said. They returned Kranz's glance. There were problems and fuel was running out. Would Armstrong and Aldrin make it? 'So now we're fighting,' Kranz said later.

They had been fighting for years, ever since Sputnik shocked the world, and even before that. Fighting the technology and the timetable. Fighting the politicians and the payroll. Fighting the Russians. Fighting to land on the Moon and be able to take off again and return to Earth.

The Apollo Moon landings were not just a technical marvel, for to think of them as only a triumph of machines and men, of combustion chambers and computers, is to diminish what was achieved. Think of them as the greatest voyages possible. Think of them as a waypoint in the evolution of our curious, explorative species. Think of them as something far, far above the normality of average human life. Think of them as a hope for our survival. Think of how they will be remembered when our Sun is dying. Think of them as a time when for a moment we achieved greatness.

All three of the crew of Apollo 11 were born in 1930. All three went into aviation and felt the sky was their natural element. Yet, having faced the danger of the unknown, 386,000 km away from Earth, they did not become friends and rarely saw one another after the mission, outside ceremonies. 'Amiable strangers' was how one of them described their relationship.

Born near Wapakoneta, Ohio, as a boy Neil Armstrong was fascinated by flying. He was always returning to his bedroom and his model aircraft. Looking back on his childhood many years later, he said he always designed his own model aircraft and never used kits. 'They had become, I suppose, almost an obsession with me,' he said. Out of school he took jobs stacking shelves at 40 cents an hour, which he put towards the $9 an hour he needed for flying lessons, and he got his pilot's licence before he could drive. It was flat country, and to him the sky seemed more important than the land. His father, Stephen Koenig Armstrong was an auditor for the Ohio state government, a stable job when the depression hit. He took Neil for an aeroplane ride when he was just ten days shy of his sixth birthday. Later Armstrong said he couldn't remember anything about it, though it obviously made a deep impression. His mother was Viola Louise Engel, a deeply religious woman, though Armstrong said she never preached to the children (who numbered three: Neil, June, and Dean). She was a very able student and was described as inventive with qualities of concentration and perseverance. It was said she longed to be a missionary and journey to distant lands. All of these qualities she passed on to her son. At seventeen he began studying aeronautical engineering at Perdue University and joined a scheme called the Holloway Plan, which was two years of study, two years of flight training and a year of service in the US Navy. He was called up to the Navy in January 1949 and during the Korean War flew 78 combat missions in Panther jets flying off the carrier *Essex*. He won three air medals. He was released from active duty in August 1952 but remained in the reserves. He had decided to become an experimental test pilot.

He never said very much. His crewmate Michael Collins said he was more thoughtful than your average test pilot, a very reserved individual. When he uttered the famous words upon stepping on to the lunar surface, few were surprised he managed to say something profound – although others were surprised he said anything at all. No wonder people called him the 'quiet aviator'. In 2000 he summed up his character: 'I am, and ever will be, a white-socks, pocket-protector, nerdy engineer, born under the second law of thermodynamics. Steeped in steam tables, in love with free-body diagrams, transformed by Laplace, and propelled by compressible flow.'

Edwin Eugene Aldrin Jr was born on 20 January 1930 in Glen Ridge, New Jersey. His parents had met in the Philippines where his father, Edwin Eugene Sr was in the Army Air Corps until he resigned on principle after his boss was court marshalled. By all accounts his father was difficult, demanding, controlling, and a major influence on Aldrin. He was the only boy in the family; hence he was called 'brother' – but his sister, Fay Ann, pronounced it 'Buzzer' and henceforth he was Buzz (he legally changed his name to Buzz in 1988). He was an athletic child and it seems he spent so much time on sports that initially he didn't get the grades to go to West Point Military

Academy. At this he knuckled down and suddenly became an 'A' student, whereupon he was accepted, graduating in 1951. Thence he went into the Air Force and became a fighter pilot in the Korean War, flying 66 combat missions. Eventually he flew F-100 Super Sabres – the first US jet aircraft that could go supersonic in level flight – out of a US air base in West Germany.

Aldrin's first wife, Joan, said of him that he was 'a curious mixture of magnificence, confidence, bordering on conceit, and humility'. He was not universally popular when he was an astronaut. Frank Borman, the commander of Apollo 8 – the first mission to travel to the Moon (on an orbital mission) – said he was worried about him before Apollo 11. 'I thought he had difficulty coping with life's simpler problems,' he said. After he returned from the Moon, fame did not wear well with him when there seemed to be none of the simpler problems, at least not for 20 years or so. The last person to walk on the Moon, Gene Cernan, said that he was called Dr Rendezvous because his thesis at MIT was on orbital rendezvous. He then added it was the only thing he could talk about, even over a cup of coffee. And then there was the squabble with Armstrong over who should step onto the Moon first …

Michael Collins, the Apollo 11 crew member who stayed in the Command Module in lunar orbit while the other two descended to the surface, was born in Rome and spent most of his childhood moving: Oklahoma, Baltimore, Ohio, Texas, Puerto Rico. His father was Major General James L. Collins, a military attaché, and Michael seemed destined for the diplomatic side of military life. But he had a streak of independence and chose the Air Force after West Point in 1952. For a decade or so he was a pilot and instructor and several times came close to working alongside Neil Armstrong. He said:

> Like most of the early astronauts, I was a test pilot, and it was a sort of step-by-step process. I went to the Military Academy. I went to West Point because it was a free and good education. I emphasize 'free'. My parents were not wealthy. When I graduated from the Military Academy, there was no Air Force Academy, but we had the chance of going into the army or the air force. [The air force] seemed like a more interesting choice. Then the question was to fly or not to fly. I decided to fly. To fly little planes or big ones? I became a fighter pilot. To keep flying the same or new ones? I became a test pilot. And so, you see, I've stair-stepped up through five or six increments then, and it was a simple, logical thing to go on to the next increment, which was higher and faster, and become an astronaut.

Of Armstrong, Collins said that he 'never transmits anything but the surface layer, and that only sparingly. I like him, but I don't know what to make of him, or how to get to know him better. He doesn't seem willing to meet anyone halfway.' He observed that among the dozen test pilots who had flown the X-15 Armstrong had been considered one of the weaker stick-and-rudder men, but the very best when it came to understanding the machine's design, and how it operated.

He described Aldrin as 'more approachable than Neil: in fact, for reasons I cannot fully explain, it is me that seems to be trying to keep him at arm's length. I have the feeling that he would probe me for weaknesses, and that makes me uncomfortable.'

But the era of these heroes in their fabulous machines was so long ago. Only 20 per cent of those alive today were around when Apollo 11 landed. Those who woke up that morning long ago feeling that the world had changed when frail humanity descended onto the Sea of Tranquillity are now an ever-diminishing minority.

The Spoils of War

There is a simple wooden house in the city of Kaluga, about 195 km southwest of Moscow, in which a deaf, self-educated Russian schoolteacher called Konstantin Eduardovich Tsiolkovsky once lived. Born in 1857, his writings, although sometimes far-fetched, put substance to mankind's nascent dreams of escaping our planet and journeying into space. His 1903 work, 'Exploration of the World Space with Reaction Machines' is regarded as the world's first scientifically viable proposal to explore space with rockets. He imagined rockets fuelled by a mixture of liquid hydrogen and liquid oxygen – the same mix used on the Space Shuttle. He developed the equation, now named after him, that provides the relationship between the changing mass of a rocket as it consumes fuel, the velocity of the exhaust gases, and the rocket's final speed. It is the foundation of astronautics. Years later he published an article on multi-stage rockets, which he said were needed to get into space. As each rocket stage used its fuel, it would break off. He predicted steering rockets, pumps to move fuel from tanks to the combustion chamber, and the need for pressurized spacesuits. Along with the later generation of rocket pioneers, the American Robert Goddard and the German Hermann Oberth, he helped prepare the way for others, and while all three dreamed of space travel, only Tsiolkovsky never thought it would come to pass. 'The Earth is the cradle of mankind, but mankind cannot stay in the cradle forever,' he wrote.

After the revolution, most were concerned with survival and there were initially few who dreamed of space travel. Yuri Vasilyevich Kondratyuk was one who did – though that was not his real name. He was born in Ukraine and while in his twenties he wrote pioneering works on rocketry such as 'The Conquest of Interplanetary Space', in which he improved on Tsiolkovsky's concepts. One of his most remarkable ideas was a mission profile for a lunar landing using two separate vehicles: a mother ship in lunar orbit, and another for the descent to and from the surface. When Americans did land on the Moon in 1969 their mission took this form. But his contribution to spaceflight was cut short. In 1916, he was conscripted into the Army to fight in Turkey. After the Bolsheviks rose to power he decided to leave the Army, but on his journey home, he was forcibly conscripted by the rebel White Army to fight against the communists. He escaped but was found by the White Army again in Kiev, whereupon after a second spell with them he deserted once more. Having fought on both sides he was in a difficult position after the revolution: both sides wanted to execute him. To save his life, his stepmother sent him the identity documents of a man named Yuri Vasilyevich Kondratyuk, who was born in 1900 and died in 1921; he assumed his new identity and tried to lead an inconspicuous life. Terrified of being found out, he did not join the burgeoning amateur rocketry groups of the 1920s and 1930s. His ideas were lost, as were his remains when he perished defending Moscow against the Nazis.

For historians, two men have come to exemplify the race to the Moon: Wernher von Braun and Sergei Korolev. They were rivals, though they never met, and von Braun knew little of Korolev except by his work. They learned, by trial and error, how to tame the explosive power of rocket fuel for a few minutes, forcing it to produce thrust. Although they lived very different lives they had many things in common: in particular a passion for spaceflight and a drive that overcame the almost overwhelming engineering and political problems they faced. They both learned to cope with failure, neglect and frustration and both in different ways carried the scars of war. Standing amid the ashes and debris of the Second World War, both looked to the Moon and felt that within their lifetimes it could be reached. But only one would live to see his dreams fulfilled.

Sergei Pavlovich Korolev was born in 1906 not far from Kiev. He came from a fractured home and was bullied at school because he was seen as a teacher's pet, due to his ability in maths. Under the care of grandparents, his family endured the many hardships that befell the people after the revolution. As a boy he was obsessed with aeroplanes and space travel, preferring his flying machines to people. When he was twenty he moved to Moscow, living in crowded squalor with his family, and attended the Bauman Higher Technical Institute where his talents came to the attention of the famed aircraft designer Andrei Tupolev, who had been one of Tsiolkovsky's pupils.

Seeking out others who shared his interests, he joined a rocket society known as GRID – the Group for the Study of Rocket Propulsion Systems. It was led by the space visionary Friedrikh Tsander who shared with Korolev dreams of flying in space. 'To Mars! Onward to Mars!' was how Tsander used to greet his fellow workers.

Tsander was born in 1887 in Riga, Latvia, where there is now a street named after him, as well as a memorial. By his twenties he wanted to make a journey into space, and in 1924 he published his landmark work entitled *Flight to Other Planets*, in which he described the design of rocket engines and made calculations of interplanetary trajectories. He had tried without success to get government support. Almost in desperation, he placed an advertisement in a Moscow newspaper calling for anyone interested in 'interplanetary communications'. Over 150 people responded. Under his leadership, GRID held public lectures and carried out small experiments in a wine cellar on 19 Sadovo-Spasskaya Street in Moscow, less than a mile from the Kremlin.

Soon Korolev replaced the ailing Tsander as leader and, with an administrative flair for which he would later become famous, established four research groups to study different rocketry problems. Now the Soviet government was impressed and soon Korolev and his colleagues were working for them. The state was already funding another small research group into solid-fuelled rockets for military use led by a young engineer called Valentin Petrovitch Glushko. He had been inspired by the works of Jules Verne and at fifteen had written a letter to Tsiolkovsky. Just three years later, in 1924, still only eighteen, he had published an article in the popular press titled 'Conquest of the Moon by the Earth'. Glushko and Korolev became friends, but that was not to last. Their difficult relationship was to be at the heart of the Soviet Space effort, becoming both its greatest strength and its greatest weakness. By the late summer of 1933 they were able to launch the Soviet Union's first liquid-fuelled rocket, powered by jellied petroleum burning in liquid oxygen. After two failures the third attempt soared to 400 metres. Korolev wrote: 'From this moment Soviet rockets should start flying above the Union of Republics. Soviet rockets must conquer space!' Tsander did not see the triumph. Five months earlier, exhausted by overwork, he had contracted typhus and died.

Wernher Magnus Maximilian Freiherr von Braun was born in Germany just before the First World War into a family that had been famous since 1245 when they defended Prussia from Mongol invasion. From an early age he showed an interest in rockets. Germany's foremost rocket scientist, Hermann Oberth, had written a book, *De Rakete zu den Planetenräumen* ('By Rocket into Planetary Space'), in which he describes a rocket equipped to go to the Moon. A young von Braun read it and was captivated. His mother gave him a telescope as a confirmation present. He made his own 'spaceship' by attaching toy rockets to a wagon and igniting them on Berlin's Tiergarten Allee.

As a young man in Hitler's Germany, von Braun took his ideas to Colonel Karl Heinrich Becker, chief of ballistics and ammunition of the Reichswehr. Becker knew von Braun's father, who was minister for agriculture. He was impressed: 'We are greatly interested in rocketry, but there are a number of defects in the manner in which your organization is going about development. For our purposes, there is too much showmanship. You would do better to concentrate on scientific data than to fire toy rockets.' In other words, he was saying to von Braun, you don't develop a weapon like this in public. Von Braun wanted to use rockets for space flight but Becker wanted a long-range missile for mass bombardment. Von Braun's friends were against the relationship with 'ignorant people who would only hinder the free development of our brainchild'. Von Braun, soon to graduate from the Berlin Institute of Technology, refused. Colonel Becker made a second offer, one that would allow von Braun to work for him *and* continue his studies. This time he accepted.

Barely twenty years old and now working in secret, von Braun joined the Army and worked under Captain Walter Dornberger on liquid-fuel rocket engines, saying later: 'We needed money for our experiments, and since the army was willing to give us help, we didn't worry overmuch about the consequences in the distant future; we were interested in one thing: the exploration of space.' To advance his career, von Braun joined the Nazi Party on 1 May 1937 and then the Waffen-SS, where he gained the rank of Sturmbannführer (major), the decoration awarded by Himmler himself. He told his colleagues that he had been conscripted, and started to lie to them about what he was really doing. After the war von Braun said: 'My refusal to join the Party would have meant I would have to abandon the work of my life.' Dornberger needed a quieter and more isolated place for the rocket tests. Von Braun remembered that his grandfather used to hunt on a pine-forested island off the Baltic coast. The sea would provide a perfect test range, so they moved to a small fishing village called Peenemunde.

While the German effort was gaining momentum, the potential of Soviet rocketry was abruptly cut short when Stalin's purges reached their inhuman climax. In the late spring of 1937 the secret police – the NKVD – arrested Marshal Tukhachevski, head of the group where Korolev was working. He was charged with having been part of an anti-Soviet, Trotskyite conspiracy. After a short trial he was executed along with his mother, sister, and two brothers.

Terrified people became informants simply to survive, among them Valentin Glushko. By the end of 1937, the secret police had Korolev and Glushko in their sights, considering them 'wreckers' of the rocket group. Glushko was arrested. Inevitably, Korolev was denounced, partly on Glushko's testimony, and was thrown into the Lubyanka – the infamous state prison in Moscow. After severe torture, he 'confessed' and was lucky not to be shot. He found himself in a cattle truck being taken three and a half thousand miles across Siberia to the Kolyma death camp, where it was said they squeezed everything out of a prisoner in the first three months because after that they didn't need him anymore.

Two chance events saved his life. A close friend, the famous pilot Valentina Grizodubova, joined with another famous Soviet aviator, Mikhail Gromov, and Korolev's mother to write a letter to the Central Committee of the Communist Party requesting a review of his case. It reached the office of Nikolai Yezhov, chairman of the secret police – who did nothing. However, his successor, the terrible Lavrenti Beriya, thought Korolev was a good example to display his leniency. The charge was altered to the less serious 'saboteur of military technology' and a new

through binoculars as the flame vanished. The rocket was over 30 km away. Walter Dornberger wrote that at this his heart was beating wildly and that he wept with joy. Later he told the engineers that they had proved that it would be possible to build piloted missiles or aircraft that could fly at supersonic speeds. 'Our rocket today reached a height of nearly 95 km. We have invaded space and shown that rocket propulsion is practical for space travel.' Dornberger thought of the possibilities of space travel but this was a time of war and the rocket was a weapon – a wonder weapon for the Third Reich. They called it the A4. Later it was renamed the V-2: V for vengeance.

The V-2 faced opposition, especially after the success of the V-1, essentially a pilotless aeroplane that ran out of fuel after about 150 miles, nose-diving to the ground with 1,800 pounds of explosives. Dornberger and von Braun were worried that money would be taken away from the V-2 in its favour. Gestapo chief Himmler, Reichsmarschall Göring and Grand Admiral Dönitz each toured Peenemunde as they tried to gain a good impression. The test prepared for Himmler spun out of control. Worried that support was ebbing, in July 1943 Dornberger and von Braun visited Hitler in his 'Wolf's Lair' in East Prussia. They carried with them scale models of rockets, drawings, diagrams, photos. Their plan worked: Hitler was supportive.

The following month, the British Air Force pounded Peenemunde with 596 heavy bombers. Many workers were killed, though only two from von Braun's inner circle. To try to avoid future attacks, production of the V-2 moved to an underground oil storage depot in the Harz Mountains near Nordhausen, leaving a relative few at Peenemunde, though it remained von Braun's base. To construct the new facility the Nazis used slave labour without mercy: the underground caverns needed enlarging and prisoners set to work with pickaxes and bare hands. Most did not survive, dying from exhaustion, starvation or execution.

Over 3,000 V-2s were launched against Allied targets. The Allies were impressed with it and as the war drew to a close it became their top priority to get their hands on it: both its technology and the engineers who built it. In a letter dated 13 July 1944, Winston Churchill requested Stalin's cooperation in locating and retrieving V-2 components that the Germans were leaving behind in their retreat. For his part, Stalin ordered the formation of a secret group to collect any rocket remains.

On 8 September 1944, two V-2s were launched from a site near The Hague in Holland intended for a site about a mile from Waterloo Station in London. One landed in Chiswick, killing three people, the youngest being three-year old Rosemary Clarke, asleep in the front bedroom of No. 1 Staveley Road. Von Braun said later: 'it behaved perfectly, but on the wrong planet.' But when he heard that the other V-2 had also hit London he drank champagne. 'Let's be honest about it,' he said. 'We were at war; although we weren't Nazis, we still had a fatherland to fight for.'

By 1945 it was clear Germany would lose the war, barring some miracle weapon – which was how many saw the V-2. Behind Hitler's back, Himmler was trying to bring all the rocket programs under his control. He tried to bribe von Braun but he was loyal to his colleagues and knew that he wouldn't live long if he was allied to Himmler against Hitler. Shortly after his refusal, von Braun was arrested on charges that he cared more about his rockets than about winning the war and had made plans to desert. It took two weeks to get him released.

It is a common belief that the impact of the V-2 on the war was limited. But things could easily have been otherwise. Eisenhower concluded that if the V-2 had come into operation just six months before it did, the invasion of Europe might not have been possible. He said: 'If they had made the Portsmouth–Southampton area one of their principal targets "Overlord" might have been written off.'

By the time von Braun was released, the Russian Army was approaching Peenemunde from the east. Von Braun later said: 'I had ten orders on my desk. Five promised death by firing squad if we moved, and five said I'd be shot if we didn't.' In mid-January 1945 he called a meeting with the other top officials at Peenemunde. The rumour was that the path of escape might be blocked very soon. Von Braun prepared to evacuate thousands of engineers, scientists and their families to central Germany. Von Braun had seen the end coming and had already started preparing his documents and equipment so that they couldn't be destroyed. He wrote on SS stationery about a fictional group he called V2BV; he said it was top secret and answerable only to Himmler. 'V2BV' was stencilled on crates of documents and equipment. General Hans Kammler ordered that he and 500 of the top scientists be separated from their families and moved to the village of Oberammergau. Von Braun feared they would be executed to deny their knowledge to the enemy. One day he pointed out to the head of the SS guard that the camp could easily be bombed by Allied aircraft. One attack could wipe out all of the Third Reich's top rocket scientists. Any guard that allowed that to happen would surely be shot. The guard agreed to let the scientists out of the camp, and to let them dress in civilian clothing so American troops would not suspect that they were of any importance.

Everyone was after the German rocket engineers and the V-2. In March the Pentagon sent a request to Colonel Holger Toftoy, chief of Army Ordnance Technical Intelligence in Europe, for a hundred operational V-2s. Toftoy sent Robert Staver to get the V-2's blueprints and documents and to find its engineers. Von Braun was top of Staver's list.

Stalin may have played a role in diverting troops towards Peenemunde rather than Berlin in the final months of the war. Just days after Hitler's suicide in Berlin, an infantry unit led by Major Anatole Vavilov from the Second Belorussian Front took control of Peenemunde. The place was deserted, with little in the way of retrievable intelligence. Stalin was furious and was reported as saying: 'This is absolutely intolerable. We defeated the Nazi armies: we occupied Berlin and Peenemunde: but the Americans got the rocket engineers. What could be more revolting and more inexcusable? How and why was this allowed to happen?'

On 2 May von Braun fled from Oberammergau. His brother, Magnus, was with him, and when they saw an approaching soldier, Magnus approached the man on a bicycle, calling out: 'My name is Magnus von Braun. My brother invented the V-2. We want to surrender.' The Americans were delighted. Operation Paperclip was the code name for the secret removal of scientists from Nazi Germany, undertaken not only for the direct benefit of the Americans but also to deny the USSR. Forty railway carriages containing the spoils – tons of documents, a hundred V-2s, test-firing rigs, a liquid oxygen plant and over 300 tons of other equipment – were dispatched to Antwerp and Navy cargo ships. Toftoy also smuggled out 118 members of the rocket team. When Churchill heard about it he was furious and complained to Eisenhower, who responded that it was too late to change things.

In June a group of Soviet engineers arrived at Peenemunde to see what they could salvage. Among them was Boris Chertok, 33, an expert on guidance systems, who immediately realized how far behind they had been. By the end of the war the most powerful operational Soviet rocket engine had a thrust of one and a half tons; the V-2 had a thrust of 27 tons. They obtained some sparse but significant items such as a combustion chamber and parts of propellant tanks, the pieces being sent back to Moscow to be examined by a group of engineers including Vasili Pavlovich Mishin, a specialist in control systems who, twenty years later, would lead the forlorn Soviet program to land a cosmonaut on the Moon.

The German rocket was far in advance of anything that the Russians, or anyone else, had. But they failed to see its full implications. They would eventually pay the price for thinking that long-range aircraft would be a superior weapon to the missile. At Peenemunde, Soviet soldiers dug out from the rubble a German edition of a book by Tsiolkovsky. On almost every page there were notes and comments made by von Braun. The Russians also found in the archives of the Nazi Air Ministry drawings of a missile designed by Soviet engineers in the late 1930s.

The Soviet Union needed rocket experts to make sense of what they were uncovering. Glushko and Korolev were recommended. Thus it was that in September Korolev, now a colonel in the Red Army, found himself in Germany. As he watched test flights of the reconstructed rocket it was clear to him that von Braun had gone further with rocket technology than anyone else – and that the Russians were going to need this technology. Korolev was never to meet von Braun; by the time he arrived in Germany, von Braun was already in America.

As the war in Europe drew to a close, Neil Armstrong was still dreaming of becoming an aircraft designer. He went to half a dozen schools as his family moved around Ohio. The war ended when he was fifteen and a year later he got his pilot's licence at the youngest age possible. His first solo flight was over his home town of Wapakoneta, landing in a grass field. Soon he would enlist in the Holloway Plan, which mapped out a program of study, flight training and naval service over the next several years. Like so many young men of his generation, he had big plans.

Object D

Just outside Huntsville in Alabama, the Americans stored their greatest prize of the Second World War – the German rocket team. A few years earlier Huntsville was small with some 13,000 inhabitants. The Army purchased 35,000 acres in 1940 to build the Redstone Arsenal and a depot for chemical weapons. Twenty thousand Army personnel arrived and thousands of construction workers. Local amenities could not keep up. By the autumn, after the typical hot and humid summer, von Braun and the first wave of his engineers were collating the V-2 documents and teaching the military what they knew about rockets. They set about assembling and launching a number of V-2s from the White Sands missile base in New Mexico.

Many did not approve of von Braun's presence in the US. In December 1946 President Truman received a letter signed by Albert Einstein and others protesting at German scientists living and working in the US. 'We hold these individuals to be potentially dangerous carriers of racial and religious hatred,' the letter read.

They were not allowed to leave their quarters without a military escort, so von Braun and his colleagues jokingly referred to themselves as 'Pops' – prisoners of peace. He was soon frustrated. His progress was not as swift as he hoped it would be.

At the end of the war the USSR may have had the most powerful land force in the world, but such forces became secondary after the bombing of Hiroshima and Nagasaki with atomic weapons. Just eighteen days after the Potsdam conference and fourteen days after Hiroshima, on 20 August 1945 a secret decree of the Soviet Central Committee and the Council of Ministers called for the formation of the Special Committee on the Atomic Bomb to direct and coordinate all efforts on the rapid development of nuclear weapons. They also needed the missiles to deliver them. Colonel General Mitrofan Nedelin and People's Commissar of Armaments Dimitri Ustinov were appointed by Stalin to lead the USSR's rocketry development. Nedelin, 44, was a brilliant officer who had used solid-fuelled Katyusha rockets during the war. Korolev was placed in charge of developing long-range missiles. His first task was to build a Soviet copy of the V-2 and improve on it, but it was clear to him that that would only serve as an interim measure. They needed better rockets of their own.

In early 1945, Mikhail Tikhonravov, who in 1933 had worked with Korolev on the development of the first Soviet liquid-fuelled rocket, brought together a group of engineers to work on a design for a high-altitude rocket to carry passengers to 190 km. Called the VR-190 proposal, it was the very first project in the Soviet Union for launching humans into space. The plan envisioned the use of a modified V-2 with a recoverable capsule for carrying two 'stratonauts'. Tikhonravov tried to obtain interest from the top:

Dear Comrade Stalin! We have developed a plan for a high-altitude Soviet rocket for lifting two humans and scientific apparatus to an altitude of 190 km. The plan is based on using equipment from the captured V2 missile, and allows for realization in the shortest time.

Stalin was interested, at least for a while, writing back, 'The proposal is interesting. Please examine for its realization.' But Tikhonravov's work stagnated. In 1947 it was renamed a 'rocket probe' and a year later a new preliminary plan was presented for approval. Further work was allowed, with one change: the launch of humans was dropped in favour of using dogs. The following year, the project was cancelled, ending the Soviet Union's first serious investigations into manned spaceflight. The issue would not re-emerge for several years.

Frustrated, Korolev took his own rocket argument to Stalin. On 14 April 1947, he was escorted into the Kremlin to meet the Soviet leader in person for the first time. 'I had been given the assignment to report to Stalin about the development of the new rocket,' Korolev later recounted. 'He listened silently at first, hardly taking his pipe out of his mouth. Sometimes he interrupted me, asking terse questions. I can't recount all the details. I could not tell whether he approved of what I was saying or not.'

By early 1948 Tikhonravov was pushing another idea – a satellite. Again he didn't receive much encouragement. That summer he read his report at the Academy of Artillery Sciences in the presence of a large group of prominent dignitaries from the military. The reaction of most was negative but Korolev was among those present and afterwards he approached his old friend, saying, 'We have some serious things to talk about.' Soon Korolev himself made plans to ask Stalin to fund the launch of an artificial satellite. The Russians now had their R-3 missile project, a rocket with a thrust of 120 tons designed to propel a three-ton warhead a distance of 3,000 km. Could it be the basis of a satellite launcher?

Meanwhile, von Braun was living in the desert at White Sands, feeling ignored and effectively doing nothing. He said, 'We can dream about rockets and the Moon until Hell freezes over. Unless the people understand it and the man who pays the bill is behind it, no dice.'

In 1950 Tikhonravov tried once again to get official interest, this time with the first detailed Soviet analysis of the requirements for launching an artificial satellite. His paper, 'On the Possibility of Achieving First Cosmic Velocity and Creating an Artificial Satellite with the Aid of a Multi-Stage Missile Using the Current Level of Technology', was presented at a special session of the Academy of Artillery Sciences. The reaction to this presentation was even worse than in 1948: some were openly hostile, some sarcastic, many silent. Korolev was one of its few supporters.

Tikhonravov wrote an article, 'Flight to the Moon', for *Pravda*, describing an interplanetary spaceship. It concluded, 'We do not have long to wait. We can assume that the bold dream of Tsiolkovsky will be realized within the next 10 to 15 years. All of you will become witnesses to this, and some of you may even be participants in as yet unprecedented journeys.' Two days later the *New York Times* said that Dr Tikhonravov left no doubt that Soviet scientific development in the field of rockets was advancing rapidly.

In the spring of 1950, a group of American scientists led by James Van Allen of Johns Hopkins University met to discuss the possibility of an international scientific program to study the upper atmosphere and outer space using rockets, balloons, and ground observations. Soon the idea expanded into a worldwide program timed to coincide with the anticipated intense solar activity from July to December 1957. They called it the International Geophysical Year (IGY). At a subsequent meeting in Rome in 1954, Soviet scientists silently witnessed the approval of an American plan to put a satellite into orbit during the IGY. In July the following year, President Dwight D. Eisenhower's press secretary James C. Hagerty said that the United States would launch 'small Earth-circling satellites'.

That same day, at the Soviet embassy in Copenhagen, Academician Sedov, Chairman of the Commission for the Promotion of Interplanetary Flights, USSR Academy of Sciences, called a press conference at which he announced: 'In my opinion, it will be possible to launch an artificial Earth satellite within the next two years.' But a sceptical Soviet leadership needed to be convinced. The same year, Korolev urged for work to begin on a satellite, but still no one was listening.

In the summer of 1951, engineers led by Korolev converged on the isolated Kapustin Yar launch site in southern Russia for the first Soviet attempt at launching a living thing into space. From an initial selection of nine dogs, two were chosen, their names Dezik and Tsygan. The launch, using the new R-4 missile, was to take place in the early morning hours so it would be illuminated by the Sun during its ascent. The launch was successful and the dogs reached a velocity of 4,200 km/h and an altitude of 101 km, officially entering space. They experienced four minutes of weightlessness. After 188 seconds the payload section separated from the main booster and went into free fall until it reached an altitude of 6 km, when the parachute deployed. Twenty minutes after lift-off, the dogs were back on the ground barking and wagging their tails – the first living things recovered after a flight into space. Two months later the United States were to achieve a similar feat.

Subsequent flights met with mixed results. Dezik and another dog, Lisa, died when their parachute failed. It was then decided that Tsygan should not fly again. Instead, in early September, engineer Anatoli Blagonravov took her back to Moscow. Russia's first canine cosmonaut lived to a grand old age. Blagonravov and the dog would often be seen walking the streets of Moscow. In total, nine dogs were flown on six launches in those early years, three of them flying twice.

When Stalin died in March 1953, it instigated the first change of leadership in the Soviet Union in more than 30 years. However, the direction of the rocketry program changed little. In early 1954, Premier Khrushchev instructed Minister Ustinov to dilute Korolev's monopoly in rocket design and construction. Ustinov came up with a plan to create two independent groups. Korolev's rival was to be the Experimental Design Bureau, formed in the Ukraine and led by 43-year-old Mikhail Yangel. It turned out to be a bad idea.

Rockets for space flight were one thing, but what mattered more to government leaders were Inter-Continental Ballistic Missiles (ICBMs). The first Soviet ballistic missile was the R-5. In February 1956, with a live atomic bomb in its nose cone, it was test launched from Kapustin Yar. Observers at the impact site in Kamchatka saw the 300-kiloton nuclear explosion. The R-5 went into service and stayed in operation for eleven years. But work was soon under way on the more powerful R-7. At last the various factors needed to put a satellite into orbit were coming together. A suitable launcher was on the horizon and Korolev's supportive colleague, Marshal Nedelin, had become Deputy Minister of Defence for Special Armaments and Reactive Technology. If a satellite were to lift off from Soviet soil, it would be Nedelin who would permit the use of a missile for such a project.

The R-7 was unlike anything created before. At launch, four conical strap-on boosters, each just over 19 metres in length, surrounded the central rocket core. It had a launch mass of 270 tons, of which about 247 tons was fuel. At lift-off, the total thrust was an impressive 398 tons. Korolev knew it could launch a satellite, if only the powers-that-be would allow it. Armed with two large sketchbooks, the ever persistent Tikhonravov made an appointment to meet Georgi Pashkov, the missile department chief at the Ministry of Medium Machine Building. One of the books contained clippings from Western magazines, including some articles by von Braun, with descriptions of American satellites. The other sketchbook contained detailed plans showing that the Russians could beat the Americans because the USSR had more powerful rockets. Pashkov was sufficiently impressed. The satellite study was approved.

Between 1950 and 1956 von Braun and his team worked for the United States' ICBM program which resulted in the Redstone rocket. He then developed the Jupiter-C – an improved Redstone. But he was frustrated. In a drawer in his desk he kept a small notebook he had had since he was sixteen. Inside were sketches for a spaceship. He had had it with him all the way from Peenemunde. But it seemed that the US government wasn't interested in space.

During his time at Huntsville von Braun became aware of the growing power of the media, especially the rise of television. In the *Huntsville Times* of 14 May 1950 there appeared a headline: 'Dr von Braun Says Rocket Flights Possible To The Moon'. He wrote a series of articles for *Colliers Magazine*, under the heading 'Man Will Conquer Space Soon'. He dreamed of 50 astronauts travelling in three huge spacecraft landing on the Moon and using the emptied cargo holds of their craft as shelters. Astronauts would drive pressurized tractors for hundreds of kilometres across the lunar surface, exploring its craters and plains. He imagined manned missions to Mars comprising a fleet of ten spacecraft, each with a mass of almost 4,000 tons, some of them carrying a 200-ton winged lander to descend to the Martian surface. To explain his vision he worked with Walt Disney on a series of films called *Man in Space* which aired in 1955.

On a tour of Korolev's rocket factory in 1956 Premier Khrushchev was ambushed when Korolev said he wanted to explain the use of his rockets for research into the upper layers of the atmosphere – at which, feeling out of his depth, the Soviet leader expressed polite interest, although it was clear by this time that most of the guests were becoming tired and bored. Detecting that his guests were in a hurry to leave, Korolev quickly moved ahead and directed everyone's attention to a model of an artificial satellite. He explained that it was possible to realize the dreams of Tsiolkovsky with the R-7 missile. He said the United States had stepped up its satellite program, but that the Soviet R-7 could significantly outdo the 'skinny' American rocket. Khrushchev began to exhibit some interest, and asked if such a plan might not harm the R-7 weapons research program. Korolev said that all the Russians would have to do was replace the warhead with a satellite. Khrushchev hesitated, suspicious of Korolev's intentions, but then said: 'If the main task doesn't suffer, do it.'

The USSR Council of Ministers issued a decree, number 149-88ss, on 30 January 1956, calling for the creation of an artificial satellite. The document approved the launch of a large satellite, designated 'Object D', in 1957, in time for the forthcoming International Geophysical Year. But by mid-1956 the Object D project was already beginning to fall behind schedule. On 14 September, the prominent mathematician Mstislav Keldysh made a personal plea to a meeting of the Academy of Sciences Presidium, saying, 'we all want our satellite to fly earlier than the Americans'.'

Indeed the Americans were getting close. Following his successful launch of the Jupiter-C in September 1956 von Braun studied charts of its flight path. It had reached an altitude of 1,120 km. He knew that if it had been fitted with a fourth stage it could have placed a satellite into orbit. It was intensely frustrating to him that he could put a satellite into orbit anytime but the government wouldn't let him. In fact, they actively prevented him from doing so, by sending observers to his rocket tests to make sure he didn't sneak one into orbit when they weren't looking. President Eisenhower and the Joint Chiefs of Staff didn't want a German – an ex-Nazi at that – to launch the first American satellite: they wanted the Navy to do it. But von Braun knew that the Navy's Vanguard rocket was inferior to the Jupiter-C, and it was behind schedule. More than once he said the Navy would lose the race to the Russians. He even said they could paint 'Vanguard' on the side of his rocket because, as he saw it, that represented their best chance of having a working rocket with that name. But no one was listening.

In the USSR the R-7 needed a new launch site. Kapustin Yar was too close to US radio monitoring sites in Turkey. The one chosen was almost in the middle of nowhere, in Kazakhstan at a place called Tyuratam. It was naked steppe, no trees, 45°C in summer, sub-zero in winter. The tsars had also used the location as a place of exile for undesirable citizens. The site was eventually to be called Baikonur.

As 1956 drew to a close Korolev was exhausted as a result of the travel from his factory at Kaliningrad to Kapustin Yar and Baikonur. Nearing breaking point, he also worried that the Americans would beat him into orbit; he had heard about the Jupiter-C launch and mistakenly believed that it had been a secret attempt to launch a satellite. He was also concerned because the results of static testing indicated that his rockets were not powerful enough for the heavy Object D satellite. He soon realized that in attempting to put into orbit a one-and-a-half-ton scientific observatory he was making things too difficult. So on 5 January 1957, Korolev sent off a letter to the government with a revised plan. He asked that permission be given to launch two small satellites, each with a mass of

40–50 kg, in the period April–June 1957, immediately prior to the beginning of the International Geophysical Year.

Fuelling for the R-7's first flight began on 15 May 1957, under the direction of Georgi Grechko, a 26-year-old engineer from Leningrad who would fly into space himself eighteen years later. The trickiest part was handling the liquid oxygen, which was at a temperature of –190°C. The process took close to five hours. When the time for the launch came, the rocket lifted gracefully into the sky but at T+98 seconds the strap-on rockets broke away from the central core and the rocket broke up.

It was one of many low points for Korolev. The 50-year-old was not in good health: he had a bad sore throat and had to take regular penicillin shots. His letters to his wife were full of doubt and frustration: 'When things are going badly, I have fewer "friends". My frame of mind is bad. I will not hide it. It is very difficult to get through our failures. There is a state of alarm and worry.'

After modifications the second R-7 rocket was taken to the launch pad in early June. There were two launch aborts traced to errors in the rocket's assembly. A third was moved to the pad for launch on 12 July. This time it lifted off into the sky but at T+33 seconds, all four strap-on boosters fell off. This was the lowest point for Korolev. There was talk of cancelling the entire program, which would end his career. He wrote to his wife: 'Things are not going very well again. Things are very, very bad.'

But the fourth R-7 launch, on 21 August 1957, was successful. The missile and its payload flew 6,500 km, the warhead finally entering the atmosphere over the target point at Kamchatka in the far east. Korolev was so excited that he stayed awake until three in the morning speaking to his deputies about the great possibilities that had opened up: the artificial satellite, and beyond that the Moon and the planets. A further test launch was successful and the satellite launch was now on. Korolev, Glushko, and the other chief designers had informally planned to stage this on the 100th anniversary of Tsiolkovsky's birth on 17 September, but that date was now unrealistic. In the meantime, Soviet spies in the United States reported that the US was ready to launch a satellite.

The next R-7 booster, this time with a satellite on board, was wheeled to the launch pad in the early morning of 3 October, escorted on foot by Korolev. He told his engineers, 'Nobody will hurry us. If you have even the tiniest doubt, we will stop the testing and make the corrections on the satellite. There is still time.' That night, huge floodlights illuminated the launch pad as the engineers in the nearby blockhouse checked the systems. History was about to be made.

The command for launch was entrusted to the hands of Boris Chekunov, a young artillery forces lieutenant. The seconds counted down to zero and Chekunov pressed the lift-off button. At exactly 22.28 Moscow time on 4 October, the engines ignited and the booster lifted off the pad. There were problems but not major ones. Satellite separation from the core stage occurred at T+324.5 seconds and the first man-made object entered orbit around the Earth. The Space Age had begun.

The Past and the Future

In the communications van, a relieved Korolev heard the *beep, beep, beep* from the orbiting satellite 101 minutes later as it completed its first orbit. They waited another orbit to tell Khrushchev. From the bunker someone called Major Dimitry Ustinov in Moscow. At just before 02.00 Moscow time, Ustinov was put through to Nikita Khrushchev, who was fast asleep.

There were four telephones in Khruschev's bedroom; the one that now rang was the large white one reserved for the most senior members of the government. When told of the success, Khrushchev said, 'Oh. Frankly, I never thought it would work' and went back to bed. His subdued reaction to the launch was not untypical; like many others, he had not grasped its importance.

Later, Korolev and a small group took off from Baikonur for Moscow. Most were exhausted and slept throughout the flight. After take-off, the pilot of the aeroplane, Tolya Yesenin, came over to talk to Korolev, telling him that 'the whole world was abuzz' with the launch. Korolev went into the pilot's cabin.

When he returned he said, 'Comrades, you can't imagine – the whole world is talking about our satellite. It seems that we have caused quite a stir.' In the morning edition of *Pravda* the news was exceptionally low key. In referring to the satellite as 'Sputnik', the Soviet media were merely using the generic Russian word for such an object (it loosely translates as 'fellow traveller'); but in doing so they conferred upon it a name that would become famous throughout the world.

A few hours after the launch, the duty officer at the CIA phoned the White House to say that the Russians had launched a satellite. President Eisenhower had left for his farm in Gettysburg to play golf. He wasn't worried. Satellites are about science, he said, not military might. A satellite did not have any effect on the United States' ability to defend itself against the Russians. He was technically right, but the public was not to believe him.

That day the Society of Experimental Test Pilots was holding a symposium in the Beverly Hilton Hotel, California. Neil Armstrong, now a young test pilot, was taking part. He was trying to find ways to get the Los Angeles press interested in the various technical presentations but they didn't care what was happening in the test-flight world. Then he heard about Sputnik. Instantly he knew it would change the world. He watched on TV as President Eisenhower completely missed the point, saying, 'What's the worry? It's just one small ball.' Perhaps, Armstrong thought, that was a facade behind which the President had substantial concerns, because if the Russians could put something into orbit, they could put a nuclear weapon anywhere in the United States. For Armstrong it was disappointing that a country that was the 'evil empire' in the minds of the American people was beating them in technology, where they believed themselves to be world leaders.

Von Braun was at a dinner party for the Defense Secretary designate Neil McElroy in the officers' mess at the Redstone Arsenal. After dinner the base public relations man ran into the room and handed von Braun a piece of paper saying that the Russians had launched a satellite. It had even been picked up by an amateur radio ham in Huntsville. An angry von Braun turned to McElroy: 'We knew they were going to do it. Vanguard will never make it. We have the hardware on the shelf. Turn us loose and we can do it. We can put up a satellite in 60 days.'

America was shocked. The seemingly unconcerned Eisenhower had misjudged the effect that Sputnik would have on the American people. A TV reporter stopped a woman in Times Square and asked for her reaction; she paused and said quietly, 'We fear this.' The next person the reporter approached said that 'somebody has fallen down on the job, badly'. Edward Teller, one of the most influential scientists in America, said the country had lost a battle more important and greater than Pearl Harbour. One astronomer said that he would not be surprised if the Russians reached the Moon within a week. Texas senator Lyndon B. Johnson, who was to play such a crucial role in getting the US to the Moon, said: 'In the open West, you learn to live closely with the sky. It is part of your life. But now somehow, in some new way, the sky seemed almost alien.'

NBC News described the beep-beep of Sputnik as the sound that forever separates the past from the future. There was a joke being told by Soviet embassy staff in Washington: over the rest of the world Sputnik's radio signal goes 'beep – beep – beep', but over the United States it changes to 'ha – ha – ha'. Such was Sputnik's impact that when radio station CKOV in British Columbia followed its special bulletin on Sputnik it played the whole of Orson Welles' *War of the Worlds* radio play without any introduction. The radio station received 60 calls saying that it wasn't aliens that had landed, it was the Russians! Elsewhere, a San Francisco reporter took the 'beat' from the emerging so-called beat generation, added the 'nik' from Sputnik, and the word 'beatnik' was born.

Khrushchev was surprised and delighted with the world-wide reaction to Sputnik. Suddenly he was enthusiastic about rockets. His son Sergei worked as an engineer for Vladimir Chelomei, one of Korolev's rivals. Sergei later said, 'He made a point of keeping up to date with the latest improvements, knowing the types of rockets and missiles that were coming into the Soviet arsenal. He took great pride that Soviet rocketry kept pace with American rocketry.' Khrushchev asked Korolev what else he could do. Korolev had an answer ready. 'We can launch a dog,' he replied.

The first Earthling to orbit the Earth was found wandering the streets of Moscow, scavenging for food in dustbins. She was a small mongrel female, malnourished but with a good person-ality; living on the streets had not brutalized her. She was about three years old and was taken in by scientist Oleg Gazenko, who called her Kudryavka or Little Curly Haired One. Originally, ten dogs were in the running to be chosen for the flight, all of them trained at the Air Force's Institute of Aviation Medicine for previous rocket flights into the upper atmosphere. They underwent many tests and procedures: subjected to noise and vibration, swung in centrifuges, kept in progressively smaller cages for up to 20 days and trained to eat high-nutrition gel. They were finally reduced to one, Kudryavka, who was now known as Laika, which means 'barker'. Air Force doctor Vladimir Yazdovsky recalls: 'Laika was a wonderful dog, quiet and very placid. I once brought her home and showed her to the children. They played with her. I wanted to do something nice for the dog. She had only a very short time to live.'

A promising 32-year-old engineer, Konstantin Feoktistov, who would later become a cosmonaut and play an important role in the design of the Salyut and Mir space stations, was placed in charge of the engineering details for the mission. As a sixteen-year-old he had acted as a scout for Soviet partisan units in his home town of Voronezh in south-western Russia. The Wehrmacht troops had used the area as a staging point for their attack on Stalingrad. Feoktistov had been captured, shot, and left for dead. But the bullet only grazed his throat, and he crawled out of a pit of corpses to reach safety under cover of darkness.

Sputnik 2 was a rushed job, put together in less than a month. All who worked on it knew it was unsatisfactory, being designed to fulfil the needs of propaganda, not science. It had a crude life-support system, food for seven days in the usual gelatinous form, and a bag to collect waste. It was so cramped there would be no room for Laika to turn around. There was no re-entry mechanism. It was to be a one-way trip. Laika was placed in the capsule three days before launch. Medical sensors were taped to her body and the readings monitored so that she did not become too distressed. Later they described her as trusting. The lift-off, on 3 November, did not go well. Although Sputnik 2 was placed into orbit, some of its thermal insulation tore loose and the temperature inside rose to above 40°C. Telemetry indicated that Laika was overheating and was agitated, though still eating. However, after five hours all went quiet.

Scientists had planned to euthanize her with poisoned food. For years afterwards they said she died when the oxygen ran out, but that wasn't true; in reality she succumbed within hours to heat stroke, an unpleasant, panicky death. Earth's first space traveller had completed 2,570 orbits by the time she was cremated on 14 April 1958 when Sputnik 2 burned up during re-entry.

In 1998 Oleg Gazenko expressed regret at the mission: 'The more time passes, the more I am sorry about it. We shouldn't have done it. We did not learn enough from the mission to justify the death of the dog.' But Korolev knew that they wouldn't stop at dogs. After winning the race to put the first artificial satellite into space they now had plans to launch the first spaceman. In his mind he had a goal of December 1960 as the earliest they could put a man into orbit. Clearly they were ahead of the US, but would it stay that way?

Things got worse for the United States. On 6 December the Navy launched their Vanguard rocket from Cape Canaveral. It reached an altitude of just over a metre, fell and exploded. The small 1.3 kg satellite was thrown clear, pathetically bleeping as it rolled away. The press called it Kaputnik. You can still see it today in the US National Air and Space Museum. As a result of this fiasco von Braun got the go-ahead to launch his small satellite, Explorer 1, which was successfully placed in orbit on 1 February 1958 by a modified Jupiter-C rocket. Von Braun knew that the United States should have been the first nation to place a satellite into orbit. Korolev knew it too. But now the space race had begun in earnest and the big question was: who would be the first to put a man in space?

Sputnik and Vanguard forced the US to reorganize and focus its efforts regarding space travel. To bring everything together Eisenhower created the National Aeronautics and Space Administration (NASA) in 1958, formed from the National Advisory Committee for Aeronautics, which had research facilities in Virginia, Ohio, and California. But NASA was meant to be bigger. It incorporated other research teams. From the Navy came the Vanguard Satellite Project team; from the Army the ballistic missile team at Redstone Arsenal, which eventually became the Marshall Space Flight Center. The Jet Propulsion Laboratory (JPL), operated by the California Institute of Technology for the Defense Department, was also made part of NASA. Within a week of NASA's inauguration in July 1958, its director T. Keith Glennan approved plans for a manned launch, giving the responsibility to the Space Task Group, a working group of NASA engineers at the Langley Research Center in Virginia, headed by Robert Gilruth. Thus Project Mercury was born. First on the list of many things to do was decide on the basic parameters of the capsule, and select some astronauts.

It was 48-year-old aeronautical engineer Harvey Allen of the Ames Laboratory who suggested a 'blunt body' design for a capsule capable of returning a man from space. He had worked on the design of ballistic missiles and was an expert in the physics of atmospheric re-entry. He showed that a blunt body, although it had greater drag than a streamlined capsule, would have a detached shock wave which would transfer far less heat to the capsule than any other design. Excessive heating was the greatest problem in the design of ballistic missiles and spacecraft, since it could melt them; the blunt body design solved this problem. Allen's idea of so-called ablative heat shields would protect the astronauts of the Mercury, Gemini and Apollo programs as their space capsules re-entered the atmosphere.

Another member of the Space Task Group was Max Faget. Eleven years Allen's junior, Faget would go on to be the major influence in the design of the Mercury capsule. He backed the blunt body concept and also showed that small retrorockets could slow a spacecraft in orbit sufficiently for re-entry, and that parachutes could be used for landing. He made preliminary sketches of a small capsule that, it was jokingly said, was meant to be worn, not flown, as it was so small. The McDonnell Aircraft Corporation won the competition to build the Mercury spacecraft. The heat shield would be a slightly convex surface of plastic and fibreglass material. The conical after-body was to be covered by shingles of high-temperature alloy similar to that used in turbine blades of jet engines. Another key design factor was the couch, that is, the astronaut's contoured seat; this would give such well-distributed support that an astronaut could withstand over 20G without injury or permanent damage.

With the design of the Mercury capsule under way, what was needed next was astronauts. In January 1959 NASA produced the blueprint for the first astronauts, on a starting salary between $8,330 and $12,770 – very reasonable for the time. They were to be test pilots with 1,500 hours experience, 25–40 years of age and no taller than five foot eleven inches. There were just under 70 applications. Nobody knew if a human could survive, let alone function, in space. To this end the doctors, according to one of them, did their best to drive the candidates crazy. By the first few days of April the final seven were told they had made it.

The 'Mercury Seven' were announced by NASA on 9 April 1959. They were Scott Carpenter, Gordon Cooper, John Glenn, Gus Grissom, Wally Schirra, Alan Shepard, and Deke Slayton. They were paraded before the press and heralded by the American public as heroes. Viewed from today's perspective the press conference was a strange event. Several of them smoked during the interviews and all, when asked, gave their home addresses. John Glenn did most of the talking, while Alan Shepard was perhaps the wittiest.

But while the selected seven began their training in the glare of the media, there is a relatively unknown back story concerning the selection of the first American astronauts. As well as the Mercury Seven there were those that were later to be called the Mercury 13, who were all women. All underwent the same physiological screening as the male astronauts, but they were not part of the official NASA selection process and none ever went into space.

It came about because one person present at the Mercury Seven press conference, Dr William Lovelace, who had developed the tests for NASA's putative astronauts, was curious to know how women would cope; so, in 1960, he invited Geraldyn (Jerrie) Cobb to take part. She was an accomplished pilot, setting world records for speed, distance and altitude in her twenties, and excelled in all the tests. Other women pilots followed; thirteen passed. They were due to go to the Naval School of Aviation Medicine in Pensacola in Florida for further testing but shortly before their visit the Navy cancelled the trip on the grounds that it was a private project and not a NASA one. Cobb wrote to President Kennedy and Vice President Johnson. Somewhat later, in 1962, hearings in the House considered the possibility of sex discrimination. John Glenn and Scott Carpenter testified that women could not be astronauts. They maintained that NASA required astronauts to be graduates of military test pilot programs and have engineering degrees – even though at the time John Glenn had no engineering degree. A year later a Soviet woman entered space. An American woman did not go into space for a further 20 years.

A few weeks after the Mercury Seven were chosen they moved to Langley Field, Virginia, where they were shown the prototype of their spacecraft. 'It didn't look very much like an airplane,' said Shepard, 'but if you were going to put a pilot in it was going to have to fly somehow like an airplane, and when you have a strange new machine, then you go to the test pilots.' At first the capsule did not have windows – engineers had thought them unnecessary as they would compromise the structural strength of the capsule. The astronauts would have none of it. They insisted on windows and on, literally, something to do. Like their Soviet counterparts, the designers wanted almost everything to be automatic, with the astronaut as a passenger who would do little except in an emergency. The future space travellers were scathing, calling this method of flying 'chimp mode'.

Sputnik shocked the US on many levels, one factor being that the Americans had no rocket that was comparable to the one that launched it. As a result von Braun was able to obtain Department of Defense authorization to develop a booster with a thrust of 680 tons. Such an unprecedented thrust was to be generated by clustering eight S-3D Rocketdyne engines, the type used in the Jupiter missiles. The program was named 'Saturn' simply because Saturn was the next outer planet after Jupiter in the Solar System.

General John B. Medaris, von Braun's boss, also wanted upper stages for the Saturn that would perform the final propulsion to send spacecraft into orbit, or beyond. Consequently the Air Force had contracted Pratt & Whitney to develop a small 15,000-pound-thrust liquid-hydrogen/liquid-oxygen engine, two of which were to power a new 'Centaur' top stage for the Air Force's Atlas rocket. This was to be the basis for the Saturn's upper stages.

After much preliminary work, NASA's Deputy Administrator Hugh Dryden announced the Apollo program to the world in July 1960. This was the United States' plan to put a man on the Moon. Long before John F. Kennedy was elected, NASA had been working on designs and guidelines for a manned Moon flight: key people would get together evenings, weekends, or whenever they could to discuss crew size and other fundamental factors. It was concluded they would need three men to do all the work, even before the landing was considered. It was decided that an oxygen atmosphere of 5 pounds per square inch (psi) was the best compromise for a system that would permit extravehicular activity without needing another module for an airlock. These guidelines, among many others, were then presented to all NASA centres and to the aerospace industry. But as yet there was no definite timescale.

Having lost the first lap in the space race to the Russians, many in the United States believed that they would win the race for the first person in space – but, for NASA, there was the difficult question of who it would be. All of the Mercury Seven astronauts were in with a chance. Alan Shepard remembers the day the choice was made:

We had been in training for about 20 months or so, toward the end of 1960, early 1961, when we all intuitively felt that Bob Gilruth had to make a decision as to who was going to make the first flight. And, when we received word that Bob wanted to see us at 5.00 in the afternoon one day in our office, we sort of felt that perhaps he had decided. There were seven of us then in one office. We had seven desks around in the hangar at Langley Field. Bob walked in, closed the door, and was very matter-of-fact as he said, 'Well, you know we've got to decide who's going to make the first flight, and I don't want to pinpoint publicly at this stage one individual. Within the organization I want everyone to know that we will designate the first flight and the second flight and the backup pilot, but beyond that we won't make any public decisions.

'So,' he said, 'Shepard gets the first flight, Grissom gets the second flight, and Glenn is the backup for both of these two sub-orbital missions. Any questions?' Absolute silence. He said, 'Thank you very much. Good luck,' turned around, and left the room.

Well, there I am looking at six faces looking at me and feeling, of course, totally elated that I had won the competition. But yet almost immediately afterwards feeling sorry for my buddies, because there they were. I mean, they were trying just as hard as I was and it was a very poignant moment because they all came over, shook my hand, and pretty soon I was the only guy left in the room.

The medical tests were unrelenting. John Glenn, who would become the first American to orbit the Earth, in the third Mercury flight, said:

They made every measurement you can possibly make on the human body, all the usual things you'd think about, plus all the other things that would occur in any natural physical exam, and then things like, oh, cold water in your ear. You sit, and you have a syringe, and you put cold water in your ear for a period of time. This starts the fluids in your inner ear, in the semi-circular canal, starts them circulating because of the temperature differential, starts them circulating, and so you get the same effect as though you'd been spun up on a chair or something like that until you are extremely dizzy, and you had nystagmus, as it's called, your eyes want to drift off. You can't keep them focused on a spot. And then they would measure how long it took for us to recover from that. There was supposed to be some correlation to something, whatever it was. They had a lot of tests like that.

No one knew how the human body would react to zero gravity. Some thought there would be only small effects; others believed that in zero G an astronaut may not be able to breathe or swallow and could be hopelessly disorientated. It was not to be. 'This is a generalization,' said Shepard, 'but it's something which I'd been doing for many, many years as a Navy pilot, as a carrier pilot; and believe me, it's a lot harder to land a jet on an aircraft carrier than it is to land a Lunar Module on the Moon. That's a piece of cake, that Moon deal! And here you had, yes, a new environment, but you know, for fighter pilots who fly upside-down a lot of the time, zero-gravity wasn't that big a deal.'

The problem turned out not to be zero G but high G. John Glenn said:

We didn't know what we'd be able to do as far as high Gs. We were going to be taking the G levels in a different direction than you normally do in a fighter aircraft. The Gs would be taken on a vector straight into your chest, because we were going to be lying down in a supine position. So you're in bed, and the whole bed was being accelerated straight upward. In other words, you're taking your Gs into your chest. We called the two different positions EI or EO – 'eyeballs in' or 'eyeballs out' – was the way the G forces were going. They didn't know what we'd be able to do at high Gs, and so they took us up to the Naval Air Development Center at Johnsville,

Pennsylvania, where they had a fifty-foot centrifuge arm, human centrifuge arm, and they had not an actual replica of the spacecraft, but a seat like you would use and a couch like you would use in a spacecraft, mounted on a capsule on the end of this fifty-foot arm, and as the arm started to rotate, went faster and faster and faster, the capsule, then, would turn so that your G vectors that you were taking in that capsule out there were the same as you'd take in a spacecraft on launch.

The Mercury 7 were not optimistic that they would all get though the program alive. Shortly after their introduction to the world they attended an Atlas launch. It exploded a minute after lift-off with Glenn remarking, 'I'm glad that one got out of the way.' Lloyds of London would not insure them. In a press conference Gordon Cooper joked, 'As the engineers say, barring any unforeseen circumstances, I'd say we've got one hundred per cent chance of success.'

In 1960, an article was published in the magazine *Missiles and Rockets*. It read:

We are in the strategic space race with the Russians, and we are losing. If a man orbits the Earth this year, his name will be Ivan. If the Russians can control space they can control the Earth, as in the past centuries the nation that controlled the seas has dominated the continents. We cannot afford to run second in this vital race. To ensure peace and freedom we must be first. Space is our great New Frontier.

The article had been ghost-written but it bore presidential campaigner J.F. Kennedy's name.

In the early summer of 1960 the capsule the Russians would be using to put a man into space was transported to Baikonur for a test launch. The spacecraft was originally designed as a camera platform but could also be used as a manned craft. For the section that carried the cosmonaut Korolev had settled on a simple yet remarkable concept: make it spherical with one side heavier with a heat shield so it would automatically turn its strengthened end forwards as it faced the friction of re-entry. Soon it was named Vostok, or 'East'. It had undergone much testing: the hatch seal was tested 50 times, spacecraft separation from the last rocket stage was tested fifteen times. The re-entry capsule was tested by dropping it from an Antonov aircraft at an altitude of 8 miles. There was, however, a problem over what to call the vehicle. Korolev proclaimed: 'There are sea ships, river ships, air ships, and now there'll be space ships!' Although the term 'space ship' was used for the first time in the official TASS news agency report, there was no indication that the mission had any relevance to a manned flight.

But all did not go well. In an orbital test the retrorocket – used to slow the capsule down so that it could start re-entry – fired on time but because the spacecraft was pointing in the wrong direction it went into a higher orbit, where it stayed for more than five years before coming back to Earth, with a piece of it striking a street in a small town in Wisconsin, USA. The problem was tracked down to a faulty sensor. A second test Vostok was prepared, with two dogs on board – Chayka and Lisichka. It was launched in July 1960, but the mission immediately ran into serious problems. Some 20 seconds after launch the rocket began to veer sideways and one of the strap-on engines exploded. The emergency escape rockets fired on top of the capsule to get the 'crew' away from the launch pad as soon as possible, but it was already too late: the dogs were dead.

Undaunted, Korolev ordered the next Vostok test mission to carry two more dogs, Belka and Strelka, along with a cargo of other biological specimens: mice, rats, insects, plants, fungi, cultures, seeds of corn, wheat, peas, onions, microbes, and strips of human skin, as well as other specimens. Lifting off on 19 August, it reached orbit but the TV pictures coming back were not good. At first, the dogs appeared still; later they became more animated, but their movements were odd and they were clearly ill. Belka squirmed and vomited. Was it possible that living things could not stand more than a single orbit in space? They parachuted in the Orsk region in Kazakhstan after a one-day, two-hour spaceflight making Belka and Strelka the first living beings to be recovered from orbit. Strelka and Belka were both taxidermied after their deaths and placed on display in the Moscow Museum of Space and Aeronautics. The spacecraft itself was only the second object retrieved from orbit: the American Discoverer 13 had pre-empted it by nine days.

It was recommended that one or two further Vostok test flights be carried out in October–November 1960, followed by two automated missions of the Vostok flight configuration in November–December. Korolev's plan was that by December 1960 the cosmonauts would be ready for a manned flight later that month, in time to beat a Mercury launch. Then disaster struck. In his Moscow office Korolev received a late-night call from Baikonur on a secure phone line informing him that there had been a major accident, the catastrophic nature of which only became clear to him through the night as more information arrived. It involved a rocket designed not by Korolev, but by his rival, Mikhail Yangel.

Yangel's group was competing with Korolev to build a new generation of ballistic missiles and had brought their first missile, the R-16, to Baikonur in mid-October for its inaugural launch. After the relative failure of Korolev's R-7 rocket (now being modified to carry the first cosmonaut) as an operational ICBM, there was a lot of pressure to bring the technically superior R-16 to operational status – it would finally justify Premier Khrushchev's bluster and bragging about Soviet rocket might. Just days before the planned launch, in a speech at the United Nations, he had boasted that strategic missiles were being produced in the USSR 'like sausages from a machine'. It wasn't true. Many important officials were at Baikonur to witness the first R-16 launch, among them Strategic Missile Forces Commander-in-Chief Nedelin who chaired the State Commission for the R-16.

Prior to the launch there had been problems with the fuelling procedures of the highly toxic propellants. A leak had caused a day's delay. On the orders of the State Commission, all repairs to the missile were carried out when it was fully fuelled – a very dangerous situation. They were completed, and 30 minutes before the launch on 24 October there were approximately 200 officers, engineers, and soldiers near the launch pad, including Marshal Nedelin, who scoffed at suggestions that he leave the area. 'What's there to be afraid of?'

he asked. 'Am I not an officer?' Yangel had gone into a bunker to smoke a last cigarette before launch. It saved his life.

An inquiry later determined that the second-stage rockets of the R-16 ignited due to a control system failure. The flames cut into the first-stage fuel tanks beneath, which exploded. Automatically activated cinema cameras filmed the explosion. People near the rocket were instantly incinerated; those farther away were burned to death or were poisoned by the toxic gases. When the engines fired, most of the personnel ran to the perimeter but were trapped by the security fence and then engulfed in the fireball of burning fuel. Deputy Chairman of the State Committee of Defence Technology, Lev Grishin, who had been standing next to Nedelin, ran across the molten tarmac and jumped on to a ramp from a height of three and a half metres, breaking both legs in the process. He died later of burns. As usual, the incident was kept secret and Marshal Nedelin was said to have died in an aircraft accident, a deception the Russians maintained until 1989. About 130 people perished as a result of the explosion, many of whom were identified only by medals on their jackets or rings on their fingers.

With the authorities keen to forget the accident, Korolev received permission to launch the fourth and fifth in the Vostok test spacecraft series. The first of these was launched without incident on 1 December 1960, into an orbit exactly mimicking the one planned at the time for a manned mission. Aboard were two dogs, Pchelka and Mushka. After about 24 hours the main engine was to fire to begin re-entry but due to a malfunction it fired for a shorter period than planned and the indications were that the landing would overshoot Soviet territory. To prevent it falling into the wrong hands, an automatic self-destruct system was triggered, blowing up the spacecraft, along with its passengers. Soon the fifth Vostok test spacecraft, carrying the dogs Kometa and Shutka, was sent on its way, but the third-stage engine prematurely cut off at 425 seconds. The emergency escape system went into operation, and the payload reached an altitude of 214 km and landed about 3,500 km downrange in one of the most remote and inaccessible areas of Siberia, the region of the Podkamennaya Tunguska River, close to the impact point of the famous 1908 Tunguska meteorite. The dogs survived. There had been two consecutive failures of Vostok test flights and it was not possible to launch a cosmonaut by February 1961. Would the Americans with their Mercury capsule win after all?

For the first group of Russian cosmonauts only men were considered. They were to be between 25 and 30 years of age, no taller than five foot nine inches and weigh no more than eleven stone four pounds. Two Air Force doctors were appointed to run the selection process and teams were sent to Air Force bases in the western Soviet Union to look for candidates. Those who got through the initial selection were interviewed but none were aware of the true nature of the mission, which was described as 'special flights'. Just over 200 passed this early screening and were then sent in groups of twenty for further testing at the Central Scientific-Research Aviation Hospital in Moscow. Testing under the 'Theme No. 6' program involved spinning the pilot in a stationary seat to test the vestibular system, and subjecting him to low pressure and increased gravity in a centrifuge. At the end of 1959 they had twenty candidates who were sent back to their units to await orders. Of the group, five were not between 25 and 30, but this condition was waived because of their performance. In the end none of them were test pilots. One of them, Vladimir Komarov, had some experience as a test engineer flying new aircraft, but the most experienced pilot, Pavel Belyayev, had accrued only 900 hours of flying time. Yuri Gagarin had flown only 230 hours.

Twelve of the twenty cosmonaut candidates undertook final medical tests. Later Gagarin recalled having a series of mathematical tests during which a voice whispered into his headphones giving him the wrong answers. 'We were tested from top to toe,' he said. The training was divided between academic disciplines and physical fitness. They had classes covering rockets, navigation, radio communications, geophysics, and astronomy. Within a few weeks each candidate had made 40–50 parachute jumps. One of the USSR's top test pilots, Mark Gallay, supervised their aircraft training. Under his direction they flew parabolic trajectories to simulate weightlessness for periods of up to 30 seconds in specially equipped aircraft. Soon they all moved to a new suburb of Moscow, about 40 km northeast, which was renamed Zelenyy ('Green'). Today it is better known by a more recent name: Zvezdny Gorodok – Star City.

Crude spacecraft simulators were installed at the new training base and because it was believed that it would be inefficient to train all twenty on one simulator, it was decided that a group of six would undergo accelerated training, and from them the selection of the first cosmonaut would be made. One of Gagarin, Kartashov, Nikolayev, Popovich, Titov, or Varlamov, would be the first. Korolev visited the centre for the first time in

June 1960, bringing with him diagrams of the Vostok capsule. The cosmonauts had only learned of his existence a few months earlier and even then he was only called the 'chief designer'.

Standing behind the cosmonauts were many young soldiers who performed much harsher tests than the prospective space travellers were subject to: stronger G forces, lower pressures, longer isolation; higher tolerance to heat, cold and toxic fumes. They were asked to take part in the space program but not to fly in space and, being patriotic, they endured with a pride that would not let them fail. They were, literally, the experimental lab rats of space. First the tests were performed on dogs; half of them died. These men were next. We do not know what the long-term effects on the soldiers was, how their suffering affected their later lives and possibly deaths, but hearsay suggests that few of them survived into the 1980s.

In January 1961 the Commission recommended the following for flights: Gagarin, Titov, Nelyubov, Nikolayev, Bykovsky, Popovich. Gagarin was the favourite. One engineer said of him, 'he would never try to ingratiate himself, nor was he ever insolent. He was born with an innate sense of tact.' Earlier the Medical Commission had described his personality: modest; embarrasses when his humour gets a little too racy; high degree of intellectual development; fantastic memory; distinguishes himself from his colleagues by his sharp and far-ranging sense of attention to his surroundings; a well-developed imagination; quick reactions; persevering; prepares himself painstakingly for his activities and training exercises; handles celestial mechanics and mathematical formulae with ease, as well as excelling in higher mathematics; does not feel constrained when he has to defend his point of view if he considers himself right; appears that he understands life better than a lot of his friends. Indeed, when the cosmonaut group had carried out an informal and anonymous survey to see who they thought should fly first, all but three of the twenty named Gagarin. Also in the running for the first flight was Gherman Titov. Twenty-five years old, he had served as a pilot in the Leningrad region. He struck many as being the most well-read of the finalists, liable to quote Pushkin or refer to Prokofiev. The last of the top three, 26-year-old Grigori Nelyubov, was perhaps the most talented and qualified of the group. He had influential supporters but was extremely outspoken.

The cosmonauts were working hard, spending long periods away from home, and when they were home they could not talk about what they did. Gagarin's wife Valya later said that if she ever asked him what he was training for he would dismiss it with a joke. Once when Yuri brought some of his cosmonaut friends home – something he rarely did – she heard them saying that soon it would either be Yuri or Gherman.

In January and February 1961, preparations for the launches of the remaining two Vostok test missions progressed. Khrushchev announced on 14 March during an interview: 'The time is not far off when the first space ship with a man on board will soar into space.'

The first man-rated Vostok spacecraft lifted off successfully just days later carrying the dog Chernushka ('Blackie'), together with mice, guinea pigs, reptiles, seeds, blood samples, cancer cells and bacteria. In the ejection seat was 'Ivan Ivanovich' – a mannequin dressed in a Sokol ('Falcon') spacesuit. The main purpose was to test radio communications with the capsule; however, they knew the Americans would be listening, and it was realized that if they heard a tape recording of a human voice, they would think it was the real thing. So it was decided to tape a popular Russian choir, and when the dummy suddenly sang like a choir, it amused all who witnessed it. The successful mission opened the way for a manned flight.

The six core cosmonauts flew to Baikonur on 17 March to witness the pre-launch operations of the final test, which went well. All re-entry procedures were conducted without any problems. Shortly afterwards a press conference was held in Moscow that relayed little information save that a successful test flight had occurred. In the audience were foreign journalists, as well as – in the front row – Gagarin, Titov, and the other cosmonauts; but of course none of the press knew that one of them would fly in space in just a few days.

Knowing that the time was nearing, Korolev, in a touching move, invited some of the original GRID veterans to his offices just a month before the first manned launch. Many of them had not seen him for years. Over vodka they spoke of the old times and their dreams of space travel. His guests knew nothing of his secret work but after they had reminisced a while he ushered them into a nearby workshop. There, in the corner, was the polished silver cockpit of the Vostok spacecraft. As they gazed at it, knowing that the age of manned space flight was about to dawn, some of them wept for joy.

Things were moving quickly. Leaving for Baikonur for the last time before the manned flight, the cosmonauts were told to tell their spouses that the launch was set for 14 April – three days later than actually intended, so they wouldn't worry as much. But who was to be the first spaceman, the cosmonaut number 1?

The State Commission had addressed the question at a meeting on 8 April. Both Gagarin and Titov had performed without fault, with Gagarin ahead in the January examinations. Nikolai Kamanin, the Air Force representative in the space program, wrote in his journal: 'Both are excellent candidates, but in the last few days I hear more and more people speak out in favour of Titov and my personal confidence in him is growing too. The only thing that keeps me from picking Titov is the need to have the stronger person for a second one-day flight.'

Photographs and details of Gagarin and Titov were sent to the Central Committee. Khrushchev replied, 'Both are excellent. Let them decide for themselves.' Finally, Kamanin, perhaps with the more arduous second flight in mind, nominated Gagarin as the primary pilot and Titov as his backup for the first flight, whose launch date was set as 11 or 12 April. Later Kamanin invited Gagarin and Titov to his office and gave them the news. Years later, when asked how he felt, Titov said it had been unpleasant. He had wanted to be the first but somehow could see why they made the choice they had. 'Yuri turned out to be the person that everyone loved. Me, they couldn't love. I'm not lovable.'

In the early morning hours of 11 April the huge doors of the main assembly building at Baikonur rolled open to reveal the R-7 booster laid on its side, borne on a converted railway carriage. Gradually, slower than walking pace, the carriage moved under the dawn sky. Alongside and watching every movement strode a nervous Korolev, who, just as with Sputnik, circled around it many times on its 4-km journey to the launch pad. This was his child. In a couple of hours it reached the pad and was levered into an upright position with an access gantry positioned alongside. Korolev would not leave it until it left the Earth. Early in the afternoon Gagarin and Titov arrived for a last-minute rehearsal. Korolev was at the point of nervous and physical exhaustion. More than once that afternoon he had to be helped to a chair for a rest. Meanwhile, an Army general at a base on the outskirts of Saratov received a phone call from the Kremlin telling him to organize the recovery of the world's first spaceman, who would be landing in his region tomorrow.

The night before launch Gagarin and Titov were assigned to a cottage near the pad area which had previously been used by Marshal Nedelin. After a light meal, they were in bed by 21.30. Korolev, unable to sleep himself, checked on them periodically. Medical sensors were attached to both cosmonauts to monitor their vital systems and strain gauges were attached to their mattresses to see how well they slept. The official history states that they both had a good night. However, Gagarin later said that he didn't sleep a wink, but worked hard to stay perfectly still lest the strain gauges on the bed would indicate he was restless and the mission be given over to Titov. Evidently Titov did the same. It was hardly the best preparation for the first manned spaceflight.

Gagarin

Pre-launch operations began in the early hours of 12 April. By dawn officials and controllers had taken up their positions beneath a crisp, cloudless sky. Gagarin and Titov were woken at 5.30 and presented with a bunch of wild flowers, a gift from the woman who had previously owned the cottage. After a short breakfast with food from a tube (meat paste, marmalade, and coffee), doctors examined them and helped them into their cumbersome Sokol spacesuits followed by a bright orange coverall. Titov was dressed first as they didn't want Gagarin to overheat. Soon they were on a bus, accompanied by eleven others, including cosmonauts Nelyubov and Nikolayev and two cameramen. The footage taken by the cameramen shows Gagarin taking his seat behind a small table. Titov can be seen walking past to take his seat further down the bus; hardly anyone seems to notice him.

At the pad, they were greeted by Korolev, Kamanin and other officials. Gagarin wanted to urinate and did so against the tyre of the bus – starting a tradition that continues to this day. After the embraces Gagarin went to the service elevator, where he halted and waved before the two-minute ride to the top. Vostok lead designer Oleg Ivanovsky helped him into the spacecraft and switched on the radio communications system. In the event of a catastrophic problem with the R-7 rocket on the pad, Gagarin was to deploy the ejection seat. However, it was clear that under these circumstances he would never gain enough altitude for the parachutes to work – so a huge net was positioned some 1,500 metres from the pad designed to catch the seat. The engineers were grateful it never had to be used.

The officials walked back to the main command bunker. There was a two-way radio and a red telephone for giving the password to fire the escape tower in case of an emergency during the first 40 seconds of the mission. The escape tower was a rocket pack connected to the capsule and should theoretically pull Gagarin free of any explosion. Only three people knew the password. Gagarin's call sign was Kedr-Cedar, while the ground call sign was Zarya-Dawn.

The flight was to be automatic. Ideally Gagarin would have nothing to do, but in the event of a malfunction he could take over control of the spacecraft by punching numbers into the keypad to release the controls, whereupon he would be able to use the thrusters to manually orientate the craft for re-entry. The numbers were in an envelope in the capsule but Oleg Ivanovsky also told him what they were: three, two and five. Yuri replied, to Oleg's surprise, that he knew: Kamanin had already told him.

Checks showed the hatch was not sealed properly. Engineers removed all 30 screws and shut the hatch a second time, whereupon all the indicators were positive. Just as they finished, the gantry started to retract automatically towards its 45-degree angle for launch – with them still on it. A frantic phone call to the blockhouse stopped the retraction for a few minutes while they descended. They finally left the vicinity about 30 minutes prior to the scheduled launch.

'Yuri, you're not getting bored there, are you?' Korolev asked.

'If there was some music, I could stand it a little better.'

'Station Zarya, this is Zarya. Fulfil Kedr's request. Give him some music. Give him some music. Did you read that?'

So many things were going through Korolev's restless mind, so many failure modes. Among the many worries, perhaps the most troubling was the prospect of the rocket's third stage failing during the ascent, depositing the Vostok spacecraft in the ocean near Cape Horn on the southern tip of Africa, an area infamous for its constant storms. Korolev had insisted that there be a telemetry system in the launch bunker to confirm that the third stage had worked as planned. If the engine worked nominally, the telemetry would print out a series of 'fives' on tape: otherwise, if it had failed there would be a series of 'twos'.

At T-15 minutes, Gagarin put on his gloves. Ten minutes later he closed his helmet. Korolev nervously took tranquillizer pills. He was well aware that of sixteen launches with this rocket, eight had failed. Of the seven Vostok spacecraft flown, two had failed to reach orbit because of booster malfunctions, while two others had failed to complete their missions. There was no American style 3, 2, 1 countdown, just a checklist, which was soon completed.

At 09.06 and 59.7 seconds on 12 April 1961, the Vostok spacecraft lifted off with its 27-year-old passenger. 'We're off,' he cried. Korolev had the abort codes in front of him in case the booster malfunctioned, but the launch trajectory was on target. At 19 seconds, the four strap-on boosters separated. The capsule's shroud broke away 50 seconds later. At about five Gs, Gagarin reported some difficulty in talking, saying that all the muscles in his face were drawn and strained. The G-load steadily increased until the central core of the launcher cut off as planned and was jettisoned at T+300 seconds. Gagarin's pulse reached a maximum of 150 beats per minute. Korolev was shaking. Incoming telemetry from the capsule began to stream a series of 'fives', indicating all was well. Then they changed to 'threes'. There were brief seconds of terror: a 'two' was a malfunction, but what was a 'three'? After a few agonising moments, the numbers reverted back to 'fives'. Feoktistov remembers that 'these interruptions, a few seconds in length, shortened the lives of the designers'.

'I see the Earth. The G-load is increasing somewhat. I feel excellent, in a good mood. I see the clouds. The landing site. It's beautiful. What beauty. How do you read me?'

'We read you well. Continue the flight.'

Orbital insertion occurred at T+676 seconds just after shut-down of the third-stage engine. The orbit was much higher than had been planned for the flight; the apogee – the furthest point from the Earth in the spacecraft's orbit – was about 70 km over the planned altitude, indicating the rocket had not performed quite as expected. Korolev had been right to worry.

Gagarin reported that he was feeling excellent. He said later:

I ate and drank normally. I could eat and drink. I noticed no physiological difficulties. The feeling of weightlessness was somewhat unfamiliar compared with Earth conditions. You feel as if you were hanging in a horizontal position in straps. You feel as if you are suspended. Later I got used to it and had no unpleasant sensations. I made entries into the logbook, reported, worked with the telegraph key. When I had meals I also had water. I let the writing pad out of my hands and it floated together with the pencil in front of me. Then when I had to write the next report. I took the pad but the pencil wasn't where it had been. It had flown off somewhere.

Once the orbit had been determined the data was sent to Moscow and reporters were instructed to open their secret envelopes. It took TASS an hour to broadcast the news:

The world's first satellite-ship 'Vostok' with a human on board was launched into an orbit about the Earth from the Soviet Union. The pilot-cosmonaut of the spaceship satellite 'Vostok' is a citizen of the Union of Soviet Socialist Republics. Major Yuri Alexeyevich Gagarin.

The Americans already knew. A radio surveillance station in Alaska had detected transmissions from the spacecraft 20 minutes after launch.

The Vostok capsule was spinning slowly and through the window Gagarin could see the black of space and the blue-white of the Earth beneath him. He could not see the stars.

The television camera trained on his face required a bright light that almost dazzled him. 'I can see the clouds, everything,' he said. 'It's beautiful.'

Over Siberia, up to the Arctic Circle, across the Kamchatka Peninsula, and into the Earth's shadow over the Pacific Vostok travelled. As its orbit took it over Cape Horn and into the South Atlantic on the final leg of its journey it was time to think about re-entry. Seventy-nine minutes after lift-off, the Vostok automatically oriented itself then, at 10.25, the retrorocket system fired for 40 seconds. As soon as the braking rocket cut out, there was a sharp jolt, and the Vostok began to rotate around very quickly. 'I had barely enough time to cover myself to protect my eyes from the Sun's rays,' said Gagarin. 'I put my legs to the window, but didn't close the blinds.'

There had been a malfunction. The large instrument section that was due to separate from the spherical descent capsule remained attached. 'I wondered what was going on and waited for the separation. There was no separation,' Gagarin said later. The mechanism released the two modules on time but the compartments remained loosely connected by a few cables. It was serious but not life-threatening. It broke off later. 'I used the telegraph key to transmit the "VN" message meaning "all goes well",' said Gagarin.

During re-entry Gagarin saw a bright purple light at the edges of the blinds and said he felt the capsule oscillate and the coating burn away with cracking sounds. He was subjected to an intense 10Gs and there were two or three seconds during which the instrument readings became blurred. 'My vision became somewhat greyish. I strained myself again. This worked.' At an altitude of 7,000 metres parachutes deployed and then the hatch was jettisoned; Gagarin was ejected and, looking down a few seconds later, he recognized he was near the Volga. He separated from his seat, and his personal parachute opened.

Ground control were frantic for several minutes when communications were cut off shortly after retrofire. Korolev telephoned Khrushchev, who was at the holiday resort at Pitsunda. 'The parachute has opened, and he's landing,' said Korolev. 'The spacecraft seems to be OK!' Khrushchev begged: 'Is he alive? Is he sending signals? Is he alive? Is he alive?'

Gagarin landed softly in a field next to a deep ravine 28 km southwest of the town of Engels in the Saratov region. He touched down just one hour and 48 minutes after launch. It took him six minutes to take off his spacesuit. 'I had to do something to send a message that I had landed normally. I climbed a small hill and saw a woman with a girl approaching me. She was about 800 metres away from me. I walked to her to ask where I could find a telephone. She told me that I could use the telephone in the field camp. I asked the woman not to let anyone touch my parachute.'

Korolev was beside himself, laughing and smiling for the first time in days. Members of the Commission flew to the landing site to inspect the capsule. Korolev did not see it until later and reportedly could not take his eyes off it, repeatedly touching it. Upon seeing Korolev, Gagarin reported quietly, 'All is well, Sergei Pavlovich.'

The *New York Times* ran the headline 'Soviet Orbits Man And Recovers Him', a headline repeated around the world. Gagarin returned to Moscow Airport flanked by an escort of fighter planes, then thousands of onlookers cheered him on a procession to Red Square, where Khrushchev, Brezhnev and other leaders of the Soviet state basked in the unqualified triumph. Derided for years by the West for its antiquated technology, the Soviet Union had taken one of the most important steps in history. But Korolev, the chief architect of this achievement, travelled several cars behind the leading motorcade and was forbidden from wearing his previous state awards on his lapel for fear that Western agents would recognize him.

However, if events had taken a different turn, it is entirely possible that Yuri Gagarin could have been the second man to go into space – although he would surely still have been the first to orbit the Earth. The first person in space could have been Alan Shepard.

Alan Bartlett Shepard Jnr, has perhaps the most remarkable story of all American astronauts. He could trace his New England ancestry back through eight generations to the *Mayflower*. Born on 18 November 1923 in Derry, New Hampshire, the son of a banker, he graduated from the US Naval Academy in 1944 and saw action in the Pacific aboard the destroyer *Cogswell*. After the war he gained his aviator's wings and went on to become a test pilot before his eventual selection as one of the Mercury Seven. Later he said, 'That little race between Gagarin and me was really, really close. Obviously, their objectives and their capabilities for orbital flight were greater than ours at that particular point. We eventually caught up and went past them, but it was the Cold War, there was a competition.'

By this time, John F. Kennedy had succeeded Eisenhower as President, elected in part because of the public's view of his predecessor's dithering in the space race – although early in JFK's presidency he too was reluctant to make a major commitment to space travel. That would change.

Shepard's flight was originally set for 24 March, but in late January the Kennedy administration received a critical report from a government advisory group known as the Wiesner Committee because it was chaired by Jerome Wiesner, Kennedy's science advisor. It said there should be an immediate delay in the first manned flight due in part to the unreliability of the Redstone booster. One of the committee's heads, George Kistiakowsky, who had been Eisenhower's science advisor, even declared that launching Shepard too early would provide the astronaut with 'the most expensive funeral man has ever had'. The Wiesner Report criticized NASA's manned space-flight program, which placed pressure on Robert Gilruth, the man in charge, as well as on NASA's new administrator, James Webb, who had succeeded T. Keith Glennan on 14 February. Consequently, they discussed the flight and told Wernher von Braun and his rocket team that a further unmanned test flight, a so-called 'booster development launch', would have to be made on the date originally set for Shepard's flight. If it was successful then he would fly on 25 April. Von Braun, who had wanted another test of the Redstone, agreed. Whatever the justification, it cost the United States the prize of sending the first person into space. Nineteen days later, up went Gagarin. The news both shattered and infuriated Alan Shepard. According to the astronauts' nurse, Dee O'Hara: 'Gagarin's flight made us look like fools. Alan was bitterly disappointed, and I could understand that.' Shepard himself said:

We had flown a chimpanzee called Ham and everything had worked perfectly except there was a relay which at the end of the powered flight was supposed to eject the escape tower, because it was no longer needed, and separate it from the Mercury capsule and eject it. For some reason with Ham's flight, it fired but it did not separate itself. So the chimp was lifted to another 25 km in altitude and another 30 km in range. There was absolutely nothing else wrong with the mission. So they said, 'Okay, let's put Shepard in the next one. Everything worked fine, so if the thing happens again, no big deal. Shepard goes a little higher.' Wernher said, 'No, we want everything absolutely right.' So we flew another unmanned mission before Gagarin flew, so it was really touch-and-go there. If we'd put me in that unmanned mission, we would have actually flown first. But it was tight.

Three weeks after Gagarin's flight Shepard finally had his chance:

The countdown had been running very, very, well. The Redstone rocket checked out well. We had virtually no problems at all and were scheduled for, I believe it was, the second of May. And, I was dressed, just about going out the door when a tremendous rainstorm, thunderstorm came over and obviously they decided to cancel it, which I was pleased they did. It was rescheduled for three days later, and of course, went through the same routine. The weather was good, and I remember driving down to the launching pad in a van which was capable of providing comfort for us with a pressure suit on and any last-minute adjustments in temperature devices and so on that had to be made; they were all equipped to do that.

We pulled up in front of the launch pad, of course, it was dark. The liquid oxygen was venting out from the Redstone. Searchlights all over the place. And I remember saying to myself, 'Well, I'm not going to see this Redstone again.' And you know, pilots love to go out and kick the tires. It was sort of like reaching out and kicking the tires on the Redstone because I stopped and looked at it, looked back and up at this beautiful rocket, and thought, 'Well, okay buster, let's go and get the job done.' So I sort of stopped and kicked the tires then went on in and on with the countdown.

There was a time during the countdown when there was a problem with the inverter in the Redstone. Gordon Cooper was the voice communicator in the blockhouse. So he called and said, 'This inverter is not working in the Redstone. They're going to pull the gantry back in, and we're going to change inverters. It's probably going to take about an hour, an hour-and-a-half.' And I said, 'Well, if that's the case then I would like to get out and relieve myself.' We had been working with a device to collect urine during the flight that worked pretty well in zero-gravity but it really didn't work very well when you're lying on your back with your feet up in the air like you were on the Redstone. And I thought my bladder was getting a little full and, if I had some time, I'd like to relieve myself. So I said, 'Gordo, would you check and see if I can get out and relieve myself quickly?' And Gordo came back. It took about three or four minutes and said, in a German accent, 'No,' he says; 'Wernher von Braun says, "The astronaut shall stay in the nosecone."' So I said, 'Well, all right that's fine but I'm going to go to the bathroom.' And they said, 'Well, you can't do that because you've got wires all over your body and we'll have short circuits.' I said, 'Don't you guys have a switch that turns off those wires?' And they said, 'Yeah, we've got a switch.' So I said, 'Please turn the switch off.' Well, I relieved myself and of course with a cotton undergarment, which we had on, it soaked up immediately in the undergarment and with 100% oxygen flowing through that spacecraft. I was totally dry by the time we launched. But somebody did say something about me being in the world's first wetback in space.

Finally he was able to say, 'Roger lift-off and the clock is started. This is Freedom 7, the fuel is go. 1.2 G, cabin at 14 psi. Oxygen is go.'

The capsule separated from the Redstone five minutes later and carried on upwards to reach an altitude of 187 km. 'OK, it's a lot smoother now,' reported Shepard. 'On the periscope: what a beautiful view. Cloud cover over Florida, three to four tenths near the eastern coast … I can see Okeechobee. Identify Andrews Island, identify the reefs.' He experienced four minutes and 45 seconds of weightlessness.

Shepard takes up the story:

We were invited back to Washington after the mission, and I got a nice little medal from the President, and which by the way he dropped. I don't know whether you remember that scene or not, but Jimmy Webb had the thing in a box and it had been loosened from its little clip, and so as the President made his speech and said, 'I now present you the medal,' and he turned around and Webb leaned forward, and the thing slid out of the box and went to the deck, and Kennedy and I both bent over for it. We almost banged heads. Kennedy made it first and he said, in his damn Yankee accent, 'Here, Shepard, I give you this medal that comes from the ground up.' Jackie Kennedy is sitting there, she's mortified and said, 'Jack, pin it on him. Pin it on him!' So he then recovered to the point where he pinned the medal on and everything was fine, and we had a big laugh out of that.

Originally Louise and I were supposed to proceed to the Congress after the White House ceremony and then have a reception, and then leave town. But Jack said, 'No, I want you to come back to the White House, have a meeting, and let's talk about your flight.' So we had the reception at the Hill, drove back, in the Oval Office there were the heads of NASA there and the heads of the government. Jack, of course, was there; and Vice President Lyndon Johnson was there. And there's a picture of me sitting on the sofa, Jack is in the rocking chair, and I'm telling him how I was flying the spacecraft, and he's leaning forward listening intently to this thing. We talked about the details of the flight, specifically how man had responded and reacted to being able to work in a space environment. And toward the end of the conversation he said to the NASA people, 'What are we doing next? What are our plans?' And they said, 'There were a couple of guys over in a corner talking about maybe going to the Moon.' He said, 'I want a briefing.'

Following Gagarin's flight, grave concern about Soviet successes had been vocalized in Congress and Kennedy asked Vice President Johnson for recommendations on activities in space that would produce 'dramatic results' and outdo the Russians. James Webb, and his deputy, Hugh Dryden, went to the White House in April 1961. The President, it was later reported, kept muttering: 'If somebody can just tell me how to catch up. Let's find somebody – anybody … There's nothing more important.' He kept saying, 'We've got to catch up.' Dryden said there was no way to catch up with the Russians when it came to orbital flights. A better idea would be a crash program on the scale of the Manhattan Project which had produced the atomic bomb. NASA had been working on something it called Apollo. It could be accelerated to put a man on the Moon within the next ten years.

But what of the Moon itself. What did we really know about it?

Mare Cognitum

Unmanned lunar missions were always part of Sergei Korolev's plans, even before he started work on Sputnik. With the success of Sputnik, resources were made available for the development of probes to the Moon and planets, but a more powerful rocket was required than that which had launched Sputnik. By the summer of 1957 the development of an addition to the R-7, called Blok E, had begun. By August 1958 both the United States and the Soviet Union were preparing to launch their first probes to the Moon. With NASA yet to be formed, the United States' Advanced Research Projects Agency – ARPA – started Operation Mona, which went under the cover name of 'Pioneer'. According to the plan, the US Air Force would use its Thor–Able rocket to launch a small probe that would enter lunar orbit.

Korolev and his engineers were preparing their E-1 lunar probe. They planned to use their new rocket to hurl a probe of about 360 kg, ten times the mass of the small US Pioneer probe, towards a lunar impact. But the rocket was giving them problems. Despite his misgivings and against his better judgement, the rumours about the imminent Pioneer launch and pressure from his superiors forced Korolev to attempt a launch. But he would have to wait: the position of the Moon at that time meant that the Pioneer launch would take place first. So on 17 August 1958 at 08.18 EDT, the Thor–Able rocket lifted off into a clear Florida sky and for the first time mankind was attempting to reach the Moon. All seemed to be going as planned but 77 seconds after launch the rocket exploded. Transmissions from the probe continued until it plummeted into the Atlantic Ocean two minutes later. It was determined that the loss was caused by the failure of a bearing in a turbo pump.

The preparations to launch the first E-1 probe early the following day were falling behind. Frustrated by a series of malfunctions on the pad, Korolev finally called off the launch after hearing of the Pioneer failure. The uncooperative rocket was returned to the assembly building; the next opportunity to launch would be in a month. On 23 September 1958 they got it back on the pad ready to try again, but soon after it took off, severe vibrations gripped its fuel tanks and 93 seconds after launch its side boosters broke loose, sending the rocket and its E-1 Moon probe back to Earth. The Russians' first attempt had ended as ingloriously as that of the Americans – but unlike their rivals their failure was kept quiet. Another rocket was hastily modified for the following month but Korolev would again have the Americans to worry about: they were preparing another Thor–Able rocket with another Pioneer probe. This time it would be under the banner of NASA, which had come into being on 1 October 1958.

As before, the United States got off the pad first. On the morning of 11 October 1958 Pioneer 1 was launched towards the Moon just seconds after the opening of its launch window. This time the 'Thor' first stage worked, allowing the high-speed 'Able' stages their chance to operate. It seemed that the probe would reach the Moon around midday two days later.

Word of the successful launch arrived in the Soviet Union just as Korolev and his team were struggling to meet their launch window the following day. But although Pioneer was the first into space, the faster trajectory of the E-1 would allow it to reach the Moon a couple of hours earlier. There was still hope. After a night of hectic preparations, the E-1 lifted off and began its chase. But it suffered so-called 'pogo' vibrations. These are caused by fuel combustion instability that leads to variation in thrust and hence oscillation along the length of the rocket, akin to a pogo stick. After just 104 seconds the rocket fell apart under the stress. It was obvious that a more thorough fix of the problem was needed and all launch attempts were put on hold. But would the American Pioneer probe reach the Moon? Fortunately for Korolev it was soon discovered that Pioneer 1 was not headed for history. A simple programming error caused the rocket's second stage to shut down early, leaving the lunar probe travelling too slowly to reach the Moon. Eventually, after a high elliptical arc, it fell to Earth. Korolev breathed a sigh of relief. The Americans, however, salvaged something from the mission. While reaching the Moon was out of the question, Pioneer 1 used its instruments to investigate the Van Allen radiation belts that circle the Earth. It showed that they faded after 15,000 km so they were not a barrier to manned lunar missions as some had feared.

If Korolev had a reprieve with the failure of Pioneer 1 it was only a brief one. The last of the original ARPA Pioneers was immediately prepared for launch in the hope of still beating the Russians, and with his rocket problems Korolev had no way to make another attempt of his own. On 7 November a Thor–Able carried Pioneer 2 into space. But while the first and second stages operated perfectly, the third stage failed and it fell back to Earth just 42 minutes after launch. NASA had one more card to play. It had a pair of small Army-developed probes that could be ready for launch in December. But by mid-November Korolev believed he had solved the pogo vibration problem and this time the laws of celestial trajectories would favour him. In December the Russians could dispatch a Moon probe two days before the United States. This time the Americans would be the ones nervously watching and waiting.

Korolev's rocket lifted off on 4 December 1958 carrying the third of the E-1 Moon probes. This time the launch was flawless, the pogo problem had been solved and Korolev dared to hope. But after four minutes the main engine spluttered and failed and he faced a tense wait to see if NASA would reach the Moon in what remained of that launch window. In the end they failed. The prize of touching the Moon had yet to be claimed.

Korolev tried again. On 2 January 1959 the Soviet Union announced the launch of Luna 1, their first acknowledged Moon probe. It was far larger than anything the United States could muster. It did not strike the Moon as intended, but passed within 6,000 km of it and went into orbit around the

Sun – whereupon the Russians renamed it Mechta, or 'Dream'. Thus the USSR added two more space firsts to their list: the first lunar flyby and the first artificial planet in the Solar System. It did produce some useful science in finding out that the Moon had no global magnetic field.

It was in September of that year that mankind finally reached the Moon for the first time. Luna 2 and its carrier rocket crashed on the edge of Mare Imbrium near the crater Archimedes, carrying two small spheres made up of pendants engraved with the Soviet coat of arms. They were designed to be scattered across the lunar landscape but if you could visit its impact site today you would see little else other than a small crater. Striking the Moon at 10,000 km an hour would have melted and vaporized them all.

It must have seemed crushing to the Americans, even more so when, less than a month later, the Russians launched the most advanced Moon probe to date and in doing so solved one of the Moon's great mysteries. On 4 October Luna 3 succeeded in returning images of its far side – the side always turned away from the Earth. It radioed back 29 very indistinct pictures that showed that the far side was very different from the near side. There was, in general, an absence of the large, dark surface regions called Mare that dominate the side always visible from Earth.

The Americans had to bear their failures and the success of the Russians while at the same time making ambitious plans for the future. It was clear to them that the Moon would be the focus for manned and unmanned exploration in the next decade. While the Russians had suffered their share of failures, their successes were spectacular and in this first round of the race to the Moon, they had clearly won. In response to the Soviet effort, the United States developed three series of unmanned spacecraft for lunar exploration: Ranger, Surveyor and Lunar Orbiter.

Ranger was designed to strike the lunar surface, taking closer and closer images before its destruction. It was hoped that the pictures would show features far smaller than could be seen with a telescope from Earth, and would tell scientists what the surface was like. Would it support a spacecraft, or would any lander sink into a sea of dust? At one time a rugged, spherical capsule capable of withstanding such an impact was incorporated into Ranger. Engineers tested a range of materials, including aluminium honeycomb, cardboard – in fact anything crushable. To their surprise, the best material by far turned out to be balsa wood. So by the summer of 1960, a 26-inch diameter sphere weighing 95 pounds was built. The plan was that after separating from the Ranger just before impact it would fall just over 300 metres to the Moon's surface, roll to a stop and begin transmitting.

But the wooden lander never got the chance to prove itself. Rangers 1 through 6, including the only three to carry the ball, all failed. Several missed the Moon because of malfunctions in the rocket launcher. Ranger 4 actually hit the Moon just over the limb. But en route a speck of foil that broke loose when it separated from its booster rocket short-circuited it. All engineers could do was track the tiny bleep from the balsa wood ball before it went behind the Moon. Premier Khrushchev was ecstatic. At a Communist Party meeting in Vladivostok he said that socialism was the only reliable launching pad for spaceships.

While analysing the Ranger 4 problem, the laboratory in charge of the project, the Jet Propulsion Laboratory (JPL) modified a Ranger to visit Venus, calling it Mariner 1: its rocket malfunctioned and it ended up at the Atlantic. Mariner 2 was sent to Venus where it became the first space probe to successfully encounter another planet, sending back valuable data. When Mariner 2 was halfway to Venus, Ranger 5 was launched towards the Ocean of Storms. It missed the Moon.

The Ranger project was now in crisis. Three commissions were established to look into the failures, one from JPL, one from NASA and one from Congress. Perhaps it was just a spell of bad luck. So just as Mariner 2 passed within 37,000 km of the Venusian cloud tops, twice as close as the Russian craft, the Ranger project was reorganized and given one last chance. So it was with high hopes that politicians and engineers watched the launch of Ranger 6 on 30 January 1964. It worked fine – except for the camera. The humiliation was complete.

James Webb, was about to cancel the project. The head of JPL, William Pickering, argued and got one final flight. At JPL's annual late-winter party, called the Miss Guided-Missile dance, Pickering was to perform the ceremony of crowning the winner of the beauty pageant. As he walked to the microphone there was spontaneous applause, 'We're going to make it work,' was all he said.

That July, Ranger 7 was heading for Mare Nubium, an area crossed by the rays from two prominent craters: Copernicus to the north and Tycho to the south. Its six cameras were each snapping five images a second. A lot could be gleaned from the pictures, which could see objects as small as half a metre in size, a thousand times better than could be obtained from the Earth. There were no jagged features, just wide-open spaces with boulders supported on the surface; and craters, craters everywhere and on all size scales. A month later the IAU recognized Ranger 7's success by renaming the impact site Mare Cognitum, the Known Sea.

The task of Ranger 8 in February 1965 was to study the central highlands looking for manned landing sites. Having gathered the data, the final Ranger – number 9 – was given to the scientists and in March 1965 its crash into the fractured crater Alphonsus was carried live on TV networks. Although there were still some who believed that a lander would sink into the lunar dust, most believed that Ranger had established the basic nature of the surface and it would most likely support the weight of a spacecraft, or an astronaut. Aeons of pitting by meteorites had not turned the surface into a deep, weak dust layer.

Meanwhile, the Russians were having mixed successes. Luna 5 was designed for a soft landing but the retrorocket system failed, and the spacecraft smashed into the Sea of Clouds. Luna 6, in June 1965, missed the Moon after a midcourse correction failed. In October, Luna 7 was intended to achieve a soft landing on the Moon but due to a premature retrofire it crashed in the Ocean of Storms. Two months later, Luna 8 was launched, again to soft-land on the Moon. This time retrofire was late, but the result was the same: the spacecraft smashed into the Ocean of Storms, which was accumulating a tidy pile of wrecked spacecraft.

Surveyor made Ranger look simple. It would have to carry out a soft landing using a solid-fuel rocket to slow its initial descent. The lander would have to aim the thrust of its braking rocket directly along its flight path to avoid tumbling. For the time it was a daunting task. Surveyor would slow to about 5 km per hour in vertical descent. To absorb an impact on hard ground, its three landing legs were fitted with shock absorbers and the footpads were made of crushable aluminium honeycomb. Its scientific team hoped that an incident during Surveyor's tests would not be prophetic. In April 1964, a test version of the spacecraft was suspended from a balloon and lifted 450 m above the New Mexico desert for a test of the landing system. Before the test could begin, a nearby electrical storm triggered the balloon's release mechanism and the test Surveyor fell to Earth and broke into pieces.

The resulting delays cost the Americans the second round in the race to the Moon, that of being the first to achieve a soft landing. On 4 February 1966, Luna 9 soft-landed, where else but in the Ocean of Storms. On the Moon four petals opened and stabilized the craft. Spring-controlled antennae popped out and a television camera looking through a rotatable mirror started taking pictures and radioing them to Earth. The pictures included views of nearby rocks and of the horizon just over a kilometre away from the spacecraft. The moonscape was like a rock-strewn desert.

As if that was not bad enough for the Americans, three months later Luna 10 became the first craft to orbit the Moon. One by one, it seemed the Russians were ticking off the major lunar accomplishments: first probe to strike the Moon; first images of the far side; first soft landing and pictures from the surface, and now the Moon's first artificial satellite. There seemed only one record left, a manned landing.

The Americans fought back and on 1 June, less than three days after a flawless launch, Surveyor 1 reached the Moon. At 80 km above its surface, its braking rocket fired for 40 seconds and to everyone's relief, Surveyor wasn't tumbling. A mission commentator called out the diminishing altitude: 1,000 feet, 500, 50, 12, then 'Touchdown'. Surveyor 1 had landed near the crater Flamsteed on the western edge of the Ocean of Storms. In Mission Control there was disbelief. Geologist Gene Shoemaker, a Ranger veteran leading one of the Surveyor science teams, recalled saying, 'My God! It landed. Hell, I wouldn't have given you a 10 per cent chance that Surveyor 1 was going to land.' Half an hour later, the first television images began to appear on the monitors at JPL, showing a round footpad perched on a dusty but firm surface. As the pictures revealed a 30 m crater rimmed with boulders, it became clear that Surveyor had been lucky to come down where it had.

The pictures were better than those from Luna 10 and for the first time you could imagine yourself standing on the Moon. There were small craters and rocks visible, and a line of boulders outlining the rim of a crater in the middle distance. When the Sun set for the fourteen-day lunar night, most thought that was the last they would be hearing of Surveyor 1 – but they were wrong: it survived five lunar nights. Surveyor 2 followed on 20 September but when one of the three rocket engines failed it started to tumble. It didn't make it, but it hardly mattered.

Vice President Johnson presented his report to Kennedy detailing what the US had to do to beat the Russians five days after Shepard's flight, calling for an acceleration of US efforts to explore space – the phrase was 'to pursue projects aimed at enhancing national prestige'. Then came Kennedy's State of the Union address to a joint session of Congress on 25 May 1961: 'I believe that this nation should commit itself to achieving the goal, before this decade is out, of landing a man on the Moon and returning him safely to Earth. No single space project in this period will be so difficult or expensive to accomplish.' Kennedy's speech was not widely reported in the Soviet media; few in their space program took any notice.

Shepard made the following analysis of it:

Just 3 weeks after that mission, 15 minutes in space, Kennedy made his announcement: 'Folks, we are going to the Moon, and we're going to do it within this decade.' After 15 minutes of space time! Now, you don't think he was excited? You don't think he was a space cadet? Absolutely, absolutely! People say 'Well, he made the announcement because he had problems with the Bay of Pigs, his popularity was going down.' Not true! When Glenn finished his mission, Glenn, Grissom, and I flew with Jack back from West Palm to Washington for Glenn's ceremony. The four of us sat in his cabin and we talked about what Gus had done, we talked about what John had done, we talked about what I had done. All the way back. People would come in with papers to be signed and he'd say 'Don't worry, we'll get to those when we get back to Washington.' The entire flight. I tell you, he was really, really a space cadet. And it's too bad he could not have lived to see his promise.

The plan was now coming together. Certain aspects of the space-craft design were already well advanced, although others would depend on the mission profile, which was yet to be confirmed. There were questions surrounding the rocket that was going to launch Apollo. As a result of numerous studies, the large rocket originally proposed by von Braun was not considered feasible so the Huntsville engineers had to scale down their ambitions. They eventually focused on a design involving five huge F-1 rocket motors on the first stage and a new hydrogen-oxygen motor for the upper stages. Calculations showed it could easily send more than 40 tons to the Moon. This was the Saturn 5.

To develop and manufacture the large Saturn stage a new plant was built at Seal Beach in California, where North American Aviation was to build major parts of the rocket. Some development and manufacture was moved into a new Douglas Center at Huntington Beach, also in California, while static testing, involving firing the rocket while held down, was carried out in Sacramento. The Marshall Center in Huntsville was also enlarged. A huge new building was erected for assembly of the first three first stages. A large stand was built to static-test the full thrust of the Saturn's five F-1 engines, which generated a staggering 180 million horsepower.

But how do you get to the Moon? What combination of rockets, docking and landers do you need? Most came to feel that lunar-orbit rendezvous was the best way to carry out the mission. With lunar-orbit rendezvous the lander leaves the mother ship in lunar orbit and goes down to the surface. Upon returning to lunar orbit, it links up with the mother ship and the astronauts transfer to it and return to Earth. It was first proposed by John Houbolt, chairman of the group that studied this plan at the Langley Research Center. It required far less mass being sent to the Moon and needed one spacecraft designed specifically for lunar landing and take-off, while the other could be designed for flying to and from the Moon and specifically for re-entry and Earth landing. If the lunar-orbit rendezvous technique was used then only one Saturn 5 launch would be needed for each Moon mission.

However, Brainerd Holmes who led Apollo in Washington was not a fan of this mission profile. He preferred Earth-orbit rendezvous, even though this would mean a dual launching of the Saturn 5 per mission, each launch carrying different space-craft components prior to their joining together in orbit. Fuel would be pumped from one to refill the other before realigning and igniting the rocket to the Moon. Although this way much larger payloads could be flown to the Moon than by a single rocket, it was complicated and risky and would probably require years to get over the technical and operational problems, thus missing the end-of-the-decade deadline.

By late 1961 the team at the new Manned Spacecraft Center were unified in support of lunar-orbit rendezvous and were investigating lunar lander design. In December of 1961 they appealed to Brainerd Holmes to approve lunar-orbit rendezvous but he was still not convinced, delaying the inevitable approval by many months. James Webb approved the lunar-orbit plan and although a few of the President's science advisors were unconvinced, the White House accepted Webb's decision. One of the NASA officials went up to Houbolt at a meeting and asked to shake the hand of the man who had saved the American taxpayer $20 billion.

So now it was known what vehicles were required. They would come to be known as the Command and Service Module, and the Lunar Module. The Apollo Command and Service Module – the CSM – was a single spacecraft, but separable into two components. The Command Module was compact and solid designed to survive the heat of re-entry after it jettisoned the Service Module and slammed into the atmosphere at 40,000 km an hour. The speed of re-entry from the Moon is nearly one and a half times as fast as returning from Earth orbit; to slow down from that speed required the dissipation of great amounts of energy. The Command Module was cone-shaped, with a blunt face for re-entry; it was 3.35 m long, 3.96 m in diameter, and weighed 6 tons.

The Service Module was the quartermaster carrying most of the stores needed for the journey: oxygen, power-generation equipment, and water as a by-product of power generation. More than that, it had a propulsion system bigger and more powerful than many upper stages of present rockets. It made all the manoeuvres needed to navigate to the Moon, to push itself and the Lunar Module into lunar orbit, and to eject itself out of orbit for the return to Earth. It was a cylinder the same diameter as the CM and 7.3 m long. Fully loaded it weighed 26 tons.

Then there was the Lunar Module, LM, pronounced 'lem' (the original abbreviation was LEM, until the word 'excursion' was dropped from the name). It only had to operate in space so its walls were flimsy and it had spindly legs. Its mission was to carry two explorers from lunar orbit to the surface of the Moon, and then send its upper half back into lunar orbit to rendezvous with the CSM. It was 7 m tall and weighed 16 tons. When Jim McDivitt returned from Apollo 9, its first manned flight, he said, 'It sure flies better than it looks.'

So these were the Apollo spacecraft: two machines, 17 tons of aluminium, steel, copper, titanium, and synthetic materials; 100,000 drawings, 26 subsystems, 678 switches, 410 circuit breakers, 65 km of wire and 33 tons of propellant.

The CSM's life-support system had remarkable efficiency and reliability. A scuba diver uses a tank of air in about 60 minutes. In Apollo it would last 15 hours. Exhaled oxygen was scrubbed to eliminate its carbon dioxide, had its moisture removed, and was reused continually. The same system kept the cabin at the right pressure, provided hot and cold water, and ran a coolant system.

Electrical power for the CM came from fuel cells and for the LM from batteries. Fuel cells used oxygen and hydrogen held as liquids at extremely cold temperatures. When combined chemically this yielded electric power and water. Storing oxygen and hydrogen required new insulated containers. It was calculated that if the Apollo hydrogen tank were filled with ice and placed in a room at 20°C, it would take eight and a half years for the ice to melt. If an automobile tyre leaked at the same rate as these tanks, it would take 30 million years to go flat.

There were 50 rocket engines on the spacecraft: sixteen on the LM, sixteen on the SM, and twelve on the CM, used to orient the craft in any desired direction. Three of the engines were much larger. All three had to work: a failure would have stranded astronauts on the Moon or in lunar orbit. So the engineers made them so simple they said that they couldn't fail.

The next thing to do was to decide where to build the moon-port. It had to be a place from which the huge rockets could be assembled and launched. The usual launch facility, Cape Canaveral, with its 7,000 hectares, wasn't large enough. Sites were considered in Hawaii, on the California coast, Cumberland Island off Georgia, Mayaguana Island in the Bahamas, Padre Island off the coast of Texas, and several others. The most advantageous site was Merritt Island, right next to the Air Force's Cape Canaveral facilities, which had been launching missiles since 1950 and NASA rockets since 1958. The site recommendation was completed July 1961. Rocco Petrone, a brilliant engineer, was in charge of the construction, having come to NASA via the Army and the Redstone rocket program. Later he would be in charge of Apollo. He spent all night printing the recommendation, and flew to Washington with Kurt Debus to brief James Webb. Debus was one of von Braun's original team. Soon 34,000 hectares of sand and scrub were acquired for NASA by the government, plus 23,000 hectares of submerged land, at a total cost of $71,872,000. Then came the construction crews, who by 1965 numbered 7,000.

At one time NASA considered preparing the Moon rocket horizontally, and then erecting it vertically on the pad as the Russians did with their considerably smaller rockets. But this would not work for such a big rocket: the Saturn 5 was to be assembled stage by stage. Weather considerations meant that an enclosed building would be essential. Even a 10- to 15-knot wind would have made outdoor assembly very difficult, and higher winds could prove disastrous. Out of these considerations came the Vertical Assembly Building, or VAB. One of the legends had it that the crane operator who set the 40-ton second stage on top of the first stage had to qualify for the job by lowering a similar weight until it touched a raw egg without cracking the shell!

But how to get the stacked Saturn 5 from the VAB to the launch pad? Early in the program it was considered moving it on its 6-km journey by water. The barge concept was familiar as the first and second stages had to come to the Cape by barge, from Louisiana and California. For this short trip why not also float the Saturn 5 and its Mobile Launcher standing upright on a barge? To see if it was feasible the Navy ran tests – which showed that it wasn't. Then engineers considered a rail system, but that was also impractical. Eventually somebody came up with the idea of using giant tracked machines like those used in strip mining. The crawler was built by the Marion Power Shovel Company, and had eight tracks, each 2 by 12.5 m, with cleats like a Sherman tank, except that each cleat weighed a ton. Mounted over these eight tracks was the platform, bigger than a baseball diamond, on which the Apollo-Saturn 5 and its Mobile Launcher would ride from VAB to pad at one mile per hour. The whole package weighed 9,000 tons, two-thirds cargo, one-third crawler.

To keep the giant rocket balanced the crawler required a levelling system that would keep it to within one degree of absolute vertical. The sensing system depended on two manometers, each 40 m long, extending like an X from corner to corner under the platform; if they showed the deck was out of level by even half an inch, it was corrected by hydraulically raising or lowering one or more of the corners. Adjustments were made many times during the trip from the VAB, especially as the crawler climbed the five-degree incline leading up to the pad.

The pads of Complex 39 became to Moon exploration what Palos was to Columbus. Pad A and Pad B were twins, each occupying about 160 acres; a Pad C had also been planned, which explains why the crawler-way from A to B had an elbow-like crook in it – the elbow would have led to Pad C. The pads were over 2 km apart so that an explosion on one would not wreck the other.

James Webb now had all the pieces in place: the Saturn 5, a Command Module for three astronauts, designed for re-entry; and a separate Service Module with a large rocket motor, attitude control jets, and fuel cells for electric power, together with fuel and oxygen. The LM would land on the Moon carrying two men to the surface and back to rendezvous with the mother ship in orbit. It was a tremendous achievement for Webb and NASA as it had been developed within one year of Kennedy's announcement. Grumman had won the contract to build the lunar lander, while North American Aviation would build the CSM.

What was needed now was a program that sat between Mercury and Apollo so that astronauts could develop the techniques needed for Apollo. This would require a spacecraft carrying an on-board propulsion system for manoeuvring, a guidance and navigation system, a radar and a controlled re-entry system. The resultant two-man Gemini spacecraft was larger than Mercury but small by Apollo standards, but the Titan II launch vehicle – the best available at that time – could not manage a larger payload.

Two months after Shepard's flight, Gus Grissom performed a similar mission in the 'Liberty Bell 7' Mercury capsule. It was straightforward, except that when he splashed down the explosive bolts on the hatch blew prematurely. There is speculation that he had accidently armed them when his elbow pressed against a switch. Grissom had to make a quick exit as Liberty Bell 7 started to sink. As he had not sealed the hose connection on his suit, it started to fill with water. The capsule was tethered to the rescue helicopter and was getting heavier. A warning light in the helicopter came on, indicating it was about to flounder. The pilot ejected the capsule, which sank thousands of metres to the sea bed. It was later discovered that the warning light was a false alarm and they could have rescued the capsule. It remained at the bottom of the Atlantic for 38 years before it was retrieved. It is now on display in the Kansas Cosmosphere and Space Center. There were no clues as to why the hatch blew.

Back in the USSR, although there were orders for the manufacture of more Vostok spacecraft, detailed plans for future missions were rather vague. Unlike the United States, which had a specific series of missions and goals as part of its Mercury project, the Soviet effort was to move forward in a rather haphazard way. It was to be their undoing. Plans for the second piloted Vostok flight focused on a day-long mission.

Titov, Gagarin's backup for the first mission, was chosen for the flight. For Titov's backup, the most likely candidate would have been Nelyubov, but Titov had been irritated by his outspoken attitude so he was dropped and Nikolayev became the back up. Three months after Gagarin's flight, Khrushchev invited Korolev and a number of other prominent space figures to meet with him on a vacation in Crimea. Korolev said that a second Vostok mission was in preparation. Khrushchev added that the launch should occur no later than 10 August. Later the reason became clear – the building of the Berlin Wall began on 13 August. Khrushchev had wanted to give the socialist world a morale boost during such a tense time.

As the launch date approached there was some trepidation because of higher than usual radiation resulting from intense solar activity, but this declined sufficiently in time for the launch. On the morning of 6 August Titov blasted off. This time the booster worked as expected but when Titov entered orbit he was not well. He felt as if he was flying upside down and in a 'strange fog', unable to read the instrument panel. On the second orbit he felt worse and thought of asking that the flight be curtailed. He tried eating a little but vomited. He carried out an experiment, manually firing the attitude control jets, but although it went well he still felt terrible, only slightly better than on previous orbits.

'Now I'm going to lie down and sleep,' he said. 'You can think what you want, but I'm going to sleep.' Flight rules said he had to keep his helmet on when sleeping but he felt that he could choke if he vomited. He rigged a piece of string to jerk open the visor in case of an emergency while sleeping. He over-slept by about 30 minutes, waking on his twelfth orbit, at the end of which he began to improve. When it came to re-entry, as with Gagarin's mission, the instrument section remained attached to the spherical descent capsule. Eventually Titov ejected after a record flight of one day, one hour and eleven minutes.

As Cosmonaut number 2 recovered from his flight, Cosmonaut number 1 was having a spot of trouble. It seems that the trials of dealing with instant fame caught up with him and he was disciplined at a Communist Party meeting on 14 November for 'acknowledged cases of excessive drinking, loose behaviour towards women, and other offences'. In what seems to have been a case of womanizing, in mid-October, Gagarin jumped out of a window of a young woman's room at a resort when his wife came knocking. He sustained a severe injury to his forehead, which left him in hospital for a while. All photos of the cosmonaut past that point show a deep scar over his left eye. Gagarin later explained to the Soviet press that he had fallen down while playing with his daughter, adding, 'it will heal, before my next space flight.' But there was never to be a next flight for him.

In November, NASA fished chimpanzee Enos from the Atlantic after he had made the second orbital test of the Mercury capsule. The third sub-orbital Redstone flight was cancelled; instead, the next one would be the more powerful Atlas booster, which would take John Glenn into orbit.

The Space Task Group was expanded into a full NASA Center and tasked with developing the spacecraft, astronaut training, and flight operations. Robert Gilruth became head of it in Houston. It was temporarily housed in about 50 rented buildings while the new Center was being designed and built. Gilruth said, 'It was a period of growth, organization, and growing pains. We were establishing new contractor relations, moving families and acquiring new homes, as well as conducting the orbital flights of Project Mercury.'

For the Soviet Union their future plans depended upon Korolev's unwritten rule that each mission be a significant advance over the previous one. A month after Titov's troubled flight, Korolev proposed a dramatic mission: three Vostok spacecraft, each with a single cosmonaut, launched on three successive days. The first pilot would conduct a three-day mission while the two others would be in space for two to three days. There would be one day when all three spacecraft would be in space. But others were not convinced and Korolev was forced to reduce the plan to two Vostok craft, to be launched by January 1962 at the earliest.

Publicity surrounding John Glenn's imminent launch did not go unnoticed in the USSR. Military-Industrial Commission Chairman Ustinov called Korolev on 7 February, just days before Glenn's flight, and ordered the dual Vostok launch in mid-March. In his diary, Kamanin commented on the stupidity of making decisions in such a way: 'This is the style of our leadership. They've been doing nothing for almost half a year and now they ask us to prepare an extremely complex mission in just ten days' time. The program of which has not even been agreed upon.' Fortunately a rocket failure at Baikonur forced a much-needed delay to the dual Vostok mission.

The Judgement of Cape Flight

In accordance with the Project Mercury tradition of including '7' in the name, John Glenn called his spacecraft 'Friendship 7', and he asked NASA artist Cecilia Bibby to paint it on the side of the capsule, allowing Bibby to become the only woman to ascend the gantry of Pad 14 at Cape Canaveral. Hearing about this, Gus Grissom dared Bibby to paint a naked woman on the spacecraft. She drew one on the inside of a cap used to cover the periscope. Glenn loved it, but Bibby almost got fired. Only the intervention of Glenn and Grissom allowed her to keep her job.

After several attempts, Glenn finally got into the capsule on 20 February 1962, intending to become the first American to orbit the Earth. The Atlas booster was unusual in its use of balloon tanks for fuel, made of very thin stainless steel with minimal rigid strength. It was the pressure in the tanks, coming from the fuel, that provided the structural rigidity required for flight. In fact an Atlas rocket would collapse under its own weight if not kept pressurized. It had to have nitrogen in the tank even when not fuelled.

'I could hear the sound of pipes whining below me as the liquid oxygen flowed into the tanks and heard a vibrant hissing noise,' Glenn said later. 'The Atlas is so tall that it sways slightly in heavy gusts of wind and, in fact, I could set the whole structure to rocking a bit by moving back and forth in the couch!'

At 09.47.39 EST, with Scott Carpenter's call of 'Godspeed, John Glenn', the rocket began to climb. 'The Atlas' thrust was barely enough to overcome its weight,' Glenn later wrote in his autobiography. 'I wasn't really off until the umbilical cord that took electrical communications to the base of the rocket pulled loose. That was my last connection with Earth. It took the two boosters and the sustainer engine three seconds of fire and thunder to lift the thing that far.'

Five minutes after the launch, Glenn reported: 'Capsule is good. Zero G and I feel fine. Oh that view is tremendous. The capsule is turning around and I can see the booster during turnaround just a couple of hundred yards behind me. It is beautiful.'

'Roger, Seven. You have a go at least seven orbits,' he was told.

Twenty-seven minutes after launch Glenn was traversing the Sahara: out of the window, he could see some long smoke trails right on the edge of the desert. Over the Indian Ocean on the first orbit he said, 'The sunset is beautiful. It went down very rapidly. I still have a brilliant blue band clear across the horizon almost covering my whole window.' After 50 minutes, concerning orbital night, 'The only unusual thing I have noticed is a rather high, what would appear to be a haze layer up some 7 or 8 degrees above the horizon on the night side. The stars I can see through it as they go down toward the real horizon, but it is a very visible single band or layer pretty well up to the normal horizon … This is Friendship Seven. I have the Pleiades in sight out here, very clear.'

Several months earlier, while getting a haircut in Cocoa Beach, Glenn had seen, in a gift shop, a little Minolta camera in a display case, which he bought for $45. NASA technicians adapted it for space and it took the most amazing images of Earth.

Glenn went on to describe something unusual he saw outside Friendship 7: 'I am in a big mass of some very small particles, they're brilliantly lit up like they're luminescent. I never saw anything like it […] they're coming by the capsule, and they look like little stars. A whole shower of them coming by. They swirl around the capsule and go in front of the window and they're all brilliantly lighted.' Later they were identified as flakes of paint from his capsule.

On the third orbit a potentially serious situation occurred. A sensor on board Friendship 7 indicated that the heat shield was loose. If it broke away during re-entry Glenn would perish.

Gene Kranz, then a Procedures Officer in Mission Control, remembers the incident as a key moment in the evolution of mission operations:

> What the crew was seeing, we were seeing on board the spacecraft in Mercury. It was only as we moved into Gemini that we recognized the need to move deeper into the spacecraft system. Part of this came about as a result of the John Glenn mission. Because in John Glenn, we

were stuck with a very difficult decision. Did his heatshield deploy or did it not? We had a single telemetry measurement that indicated that the heatshield had come loose from the spacecraft. Now, if we believed that measurement and the heatshield had come loose, we had one set of decisions that involved sticking our neck out by retaining the retrorocket package attached during the entry phase. We didn't know whether it would damage the heatshield. We didn't know whether we had sufficient attitude control authority. So if the heatshield had come loose and we believed that measurement, we'd go that direction. But if the heatshield had not come loose, that measurement was wrong and we wouldn't do anything different. So it was a very difficult decision.

I remember this one very clearly, because the engineers would come and say, 'Nah, the heatshield can't have come loose!' And Chris Kraft [Mercury Flight Director] would look at them, and he'd say, 'Well, how about this measurement we're seeing? What's the worst thing that would happen if it had come loose?' And they'd always end up in a position that says, 'Well, maybe John Glenn isn't going to make it home.' 'Well then, what are we going to do about it?' So, because of Kraft, this entire business of ground control, I think, really came into being on the Mercury–Atlas 5.

Apparently we had a bad sensor on that flight, and we thought that the heat shield had been released. Well, the arrangement was such that if the heat shield were released, the straps would have still held it on. So everybody was concerned that the heat shield had been released, because that's what the instrumentation said. Well, they called me up – I was back in Houston – and asked me what about it, and I said, 'Well, you can enter that way because we've got wind tunnel data that said the thing will be stable,' and they did.

Now they had to tell John Glenn about the problem. But they didn't tell him the whole story. Schirra said, 'This is Texas Capcom, Friendship Seven. We are recommending that you leave the retropackage on through the entire re-entry.'

'This is Friendship Seven. What is the reason for this? Do you have any reason? Over.'

'None at this time. This is the judgement of Cape Flight.'

Six minutes before retrofire, Glenn manoeuvred Friendship 7 into a 14-degree, nose-up attitude. The first retrorocket fired. The second and third retrorocket firings came at five-second intervals, slowing the capsule sufficiently to drop it out of orbit. Once again, Schirra said, 'Keep your retro pack on until you pass Texas.'

'I looked around the room,' wrote Gene Kranz in his autobiography, *Failure Is Not an Option*, 'and saw faces drained of blood. John Glenn's life was in peril.'

In Florida, Capcom Al Shepard said, 'We feel it is possible to re-enter with the retro package on. We see no difficulty at this time with this type of re-entry.'

After the mission Glenn said he was annoyed at being kept in the dark about such a serious problem. During re-entry the retrorocket package's three metal straps melted. A fragment struck the window. Later Glenn said he expected to feel the heat growing on his back. He did not know if the pieces that were coming away were pieces of the retrorocket package or the heat shield. 'My condition is good, but that was a real fireball. Boy. I had great chunks of that retropack breaking off all the way through.'

Glenn became a national hero and had one of the biggest New York tickertape welcomes ever. But it seemed that the flight of Friendship 7 would be his last.

Well, after my flight I wanted to get back in rotation and go up again. Bob Gilruth, who was running the program at that time, said that he wanted me to go into some areas of management of training and so on, and I said I didn't want to do that. I wanted to get back in line again for another flight. But he said headquarters wanted it that way, at least for a little while. And I didn't know what the reason for this was, and I kept going back. Every month or two I'd go back and talk to him again about when do I get back in rotation again, and he'd tell me, 'Well, not now. Headquarters doesn't want you to do this yet.'

I don't know whether he was afraid of the political fallout or what would happen if I got bagged on another flight. I don't know what it was, but apparently he didn't want me used again right away. So that's the reason I never got another flight. Bob Gilruth kept saying, well, that he wanted me here as a training and management, plus the upcoming flights should go to people who would be useful for the early lunar landings, and that by the time those

were expected to occur, I'd be over fifty, and that might
be a little too old.

By a curious twist of fate John Glenn did get into space again –
36 years later! It was on the Space Shuttle and Glenn, then aged
77, conducted tests to see the difference between his adaptation
to zero gravity and his younger crew members.

Deke Slayton was the obvious choice for the second orbital
Mercury mission but doctors discovered he had a slightly irregu-
lar heartbeat, so he was grounded. The rest of the Mercury
crew appealed to President Kennedy to overrule the doctors;
Kennedy assigned this hot potato to his Vice President, who
invited the astronauts, along with Gilruth, to a weekend at
the LBJ ranch to thrash out this complaint and others. Deke
remained grounded but his comrades elected him their leader,
thereby conferring on him (with Gilruth's approval) the power
that would control the destinies of all astronauts for the next
decade.

Slayton's back up was Wally Schirra, but instead the mission went to Scott Carpenter, John Glenn's back up for the
first US orbital flight. He went into space in May. His flight, in
'Aurora 7', has been unfairly criticized as a poor flight. It was
planned to be a scientific mission as well as conducting the most
thorough workout of the Mercury capsule to date. However,
various malfunctions resulted in the need for frequent attitude
corrections and a heavy use of fuel. Carpenter also made a serious error in not turning off the automatic orientation system
when he switched to manual control prior to re-entry; the result
was a critical waste of fuel. Further malfunctions and the low

level of fuel meant that Carpenter did not align Aurora 7 correctly for retrofire, which he also activated five seconds too late. Re-entry itself was dramatic with the capsule oscillating wildly. It overshot the landing zone by 400 km. Carpenter eventually left NASA.

Meanwhile, the Russians had approved the hiring of 60 new cosmonaut trainees including five women: Tatyana D. Kuznetsova, 20; Valentina L. Ponomareva, 28; Irina B. Solovyeva, 24; Valentina V. Tereshkova, 24; and Zhanna D. Yerkina, 22. Solovyeva had 900 parachute jumps to her credit, followed by Tereshkova with 78 and Ponomareva ten. Although Ponomareva was clearly the most accomplished pilot, Gagarin opposed her inclusion because she was a mother. Tereshkova, did not have any academic honours but had been an active member of the local Young Communist League.

Cosmonauts Nikolayev and Popovich were the obvious candidates for the two missions. One of the few bachelors in the team, 32-year-old Nikolayev began his career as a lumberjack before joining the Soviet Air Force, receiving his pilot's wings in 1954. Popovich, 31, had had a distinguished career in the Soviet Air Force before receiving the Order of the Red Star for an assignment in the Arctic. His wife Marina was one of the most accomplished female test pilots in the USSR.

On 11 August Nikolayev took off. Korolev was nervous throughout the ascent phase and held tightly to the red telephone with which he would give the order to abort the mission in case of a booster failure. Khrushchev spoke to Nikolayev four hours into the mission, and the world saw Nikolayev smile on TV. As Vostok 3 passed over Baikonur at 11.02 a day afterwards,

Vostok 4 climbed after it. It was the first time that more than one piloted spacecraft, or indeed more than one human, had been in orbit. Western media was surprised by the second launch, speculating that there would be a docking. There was talk that the mission was a rehearsal for a Moon flight but careful commentators noticed that this was not a true rendezvous, just two spacecraft launched into similar orbits neither of which could alter them. Both Vostoks fired their retrorockets within six minutes of each other on 15 August. Nikolayev landed after a three-day, 22-hour, 22-minute flight, during which he had circled Earth 64 times. Popovich landed 200 km away after a two-day, 22-hour, 57-minute flight and 48 orbits.

Korolev breathed a sigh of relief; his political masters were satisfied. But his health was worsening. He had been in poor condition for years: the privations of the labour camps had never left him. His busy work schedule aggravated matters; it was common for him to work eighteen hours a day for several weeks. He found it hard to delegate, often involving himself in trivial matters he should have left to others. Soon after the return of the twin Vostoks he experienced intestinal bleeding. After a stay in the hospital, he was ordered to take a holiday at the seaside resort of Sochi – but he took his work with him and was constantly on the telephone.

NASA needed new astronauts, the ones who would undertake the Moon missions. In September 1962 Tom Stafford met Pete Conrad and John Young in the lobby of the Rice Hotel in Texas. All were Navy men. They had all been given instructions to fly to Houston Hobby Airport. There, NASA security officers would pick them up and take them downtown to the hotel, where they were all registered under the name of the manager, Max Peck. This was to confuse reporters who were on the prowl. The three had a few beers and noted that Jim Lovell and Ed White had been seen in the hotel. Conrad and Lovell had actually been in the running to be finalists for the Mercury program but had been dropped for some minor medical issue. Soon Frank Borman and Jim McDivitt arrived as well, and Elliot See, one of two civilians selected. The other was Neil Armstrong.

On the morning of 16 September the new recruits travelled to Ellington Air Force Base to meet Deke Slayton and Al Shepard, along with the director of the Manned Spacecraft Center Robert Gilruth, Walt Williams, head of flight operations, and Shorty Powers, the public affairs officer 'There'll be plenty of missions for all of you,' said Gilruth. Deke Slayton spoke about the new pressures they would all face, especially business dealings and freebies. 'With regard to gratuities,' Deke said, 'If there is any question, just follow the old test pilot's creed: Anything you can eat, drink, or screw within twenty-four hours is perfectly acceptable.' Gilruth was said to have blushed at this and Walt Williams choked, holding up his hand, 'Within reason, within reason.' But no one was thinking of the goodies that day; what they were thinking was, *which one of us is going to be first on the Moon?*

Once they had settled into their training they were given technical assignments – areas of the Gemini program that they were to monitor and take part in the development of. Frank Borman was assigned to the Saturn rocket; Jim McDivitt was allocated guidance and control; John Young got environmental controls and pressure suits; Jim Lovell got recovery and re-entry matters; Elliot See got electrical systems and mission planning; Pete Conrad would oversee cockpit layout, and Neil Armstrong got trainers and simulators.

The Mercury program was gaining momentum with the launch of Walter Schirra in the 'Sigma 7' capsule in October 1962. He intended to fly a technically perfect mission. He said:

Not to criticize John and Scott, but the mission was designed to have a chimpanzee in there. They replaced the chimp. But that meant they had to have a lot of automatic manoeuvres. Automatic manoeuvres took a tremendous amount of attitude control fuel. I said, 'I don't want to do that. I just want to save that.' And as a result, I ended up, I think, about retrofire, about 80% of my attitude fuel was still remaining.

Schirra talked about 'fireflies' – a phenomenon that Glenn had observed:

As water came out of the spacecraft it froze instantaneously into one snowflake, but a very tiny, tiny snowflake. These stuck on the outside of the spacecraft. They drifted around. This was what John called fireflies, is what Scott got involved with banging the spacecraft and watching them come off. And as a result, both of them lost sight of the fact they had to have fuel enough to fly the mission. John got a little wrapped up; I did, too, because I was his Capcom in California, on the retro-rocket package that had to be kept on because of a false signal that said his heatshield had detached, when in fact it turned out it had not. But at any rate, that became kind of a traumatic part of John's mission. But in both cases, they almost ran out of attitude control fuel; and that kind of shook me up, because there's no reason to do that. In fact, I alienated some of the flight controllers because, after drifting for a while, I put it back into automatic control: 'I'm in chimp mode now'; it didn't go over too well.

On 2 November President Kennedy met with space officials at the White House. A recording of it exists in the J.F. Kennedy Presidential Library. His science advisor Jerome Wiesner said, 'We don't know a damn thing about the surface of the Moon. And we're making some wild guesses about how we're going to land on the Moon and we could get a terrible disaster from putting something down on the Moon.' Kennedy took this in but he was thinking of the international situation: 'Everything that we do ought to really be tied into getting onto the Moon ahead of the Russians.' Webb didn't want this, he wanted a NASA that was more than a Moon program: 'Why can't it be tied to pre-eminence in space, which are your own—' Kennedy interjected, 'Because, by God, we keep – we've been telling everybody we're preeminent in space for five years and nobody believes it because they have the booster and the satellite.' Later Webb returned to the point: 'I think it is *one* of the top-priority programs.' Kennedy let him speak, then again made his original point. 'Jim, I think it is top priority. I think we ought to have that very clear … this is, whether we like it or not, in a sense a race. If we get second to the Moon, it's nice, but it's like being second any time.'

On 27 March 1963, three unflown cosmonauts, Nelyubov, Anikeyev, and Filatev, were returning to their training centre after an evening in Moscow. They had been drinking and became involved in an altercation with a military patrol at a railway station. Nelyubov threatened to go over the head of the offended officers if they filed a formal report against the three of them. Later officials at the Cosmonaut Training Centre requested that the duty officer not file a report against the three. He agreed, provided they apologize for their behaviour. Although Anikeyev and Filatev agreed to make peace, Nelyubov refused so the offended duty officer filed a report against the three of them, and within a week all three were dismissed from the cosmonaut team. Nelyubov was one of the brightest and most qualified cosmonauts, he had served as Gagarin's second backup during the first Vostok mission, and he certainly would have gone into space soon.

There was some discussion among Kamanin and the cosmonauts in later months about bringing Nelyubov back. However, presumably facing the prospect of never going into space, Nelyubov suffered from a psychological crisis made worse as cosmonauts who were junior to him started flying their space missions. By 1966, he was despondent. The final Air Force report on him states: 'On February 18th while in a state of drunkenness, he was killed by a passing train on a railroad bridge at Ippolitovka station on the Far Eastern Railroad.' He was 31 years old.

The final Mercury flight occurred in May 1963 when the relaxed Gordon Cooper flew in his capsule, 'Faith 7'. He had to put up with what was becoming a common problem on Mercury flights – spacesuit overheating. The flight went well and he remarked at how much detail he could see down on the ground. He deployed a flashing beacon from the nose of his capsule to test how far he could see it as it drifted away, important for future rendezvous missions. Then there was trouble, as he later described:

On the 19th orbit a warning light came on. The .05 G green warning light came on, which is the light that tells you you're starting to re-enter. I was sure that I wasn't re-entering, because there had been nothing to slow down my speed at all. And, of course, as usually happened on these missions, we had long spaces that we were out of radio contact; and I was out of radio contact when this happened. So when I got in radio contact first time, the Cape was kind of concerned when they heard about this light on. Then we proceeded on the next orbit or so to try to analyse, go through various procedures to try to find what it was. And we realized I was, slowly but surely, having an electrical fire from my relays; and they did short out the inverters. So, eventually I lost my total electrical system.

'It meant that I had the manual push/pull rods to activate the jets for attitude control. I had eyeballs out the window for my attitude – my pitch, roll, and yaw attitude. I had a wristwatch for timing. And I had to activate each and every one of the relays, and I'd have to manually fire the retros while manually flying the spacecraft. So, everything had to be done manually. I'd have to control the spacecraft all the way through re-entry. I'd have to put my drogue out manually. And I'd have to deploy my parachute manually. I'd have to deploy the landing bag manually.

In the end it was a perfect splashdown just 7,000 yards from the USS *Kearsarge* in the Pacific.

Mercury ended with a total of two days, five hours and 55 minutes cumulative space time from six missions. It might not have sounded much but it was a sound start, verifying the technology necessary to maintain a human in Earth orbit for a short period of time.

A Fair Solar Wind

How was the USSR to respond to the impressive Mercury flights and demonstrate its own superiority? It was decided that the next flight of the Vostok was to include a woman. Reports on the candidates stated that Valentina Ponomareva had the most thorough preparation and was more talented than the others:

> she exceeds all the rest in flight, but she needs a lot of reform as she is arrogant, self-centred, exaggerates her abilities and does not stay away from drinking and smoking. Solovyeva is the most objective of all, more physically and morally sturdy, but she is a little closed off and is insufficiently active in social work, Tereshkova is active in society, is especially well in appearance, makes use of her great authority among everyone who she knows. Yerkina has prepared less than well in technical and physical qualities, but she is persistently improving and undoubtedly she will be a rather good cosmonaut.

The report then reached a conclusion: 'We must first send Tereshkova into space flight, and her double will be Solovyeva.' It was said among the trainees that Tereshkova was Gagarin in a skirt! The flight was set for August 1963 but almost immediately Korolev discovered a problem. His engineers realized that the operational lifetime of both the slated spacecraft was due to expire in May–June 1963, well before the August flight, and there was no possibility of extending their 'shelf life' – they had to launch them or scrap them. So they changed the timetable. The first spaceship, in May or June, would carry a man into orbit for a full eight days, while the second would carry the first woman into space for two to three days. The choice for the first mission was Bykovsky.

In her book, *The Female Face of the Cosmos*, Ponomareva later wrote:

Korolev … asked why I was sad and whether I would resent it if I do not fly. I rose and said with emphasis: 'Yes, Sergei Pavlovich, I would resent it very much! Pointing his index finger at me, Korolev said: 'You are right, you fine girl, I would have resented it too.' He spoke with emphasis, very emotionally. Then he has kept silent for a while, gave every one of us a long attentive look, and said: 'It's all right, you'll all fly into space.' The session of the State Commission on 21 May, 1963 was short, and there was no miracle. It was announced that Valentina Tereshkova was appointed the commander of the space ship, and Irina Solovyeva and Valentina Ponomareva were the back-ups. As I remember the physician Karpov's explanation, two back-ups, instead

of one as for men, were appointed 'with the consideration
of the complexity of the female organism.'

Ponomareva would have made a good cosmonaut. She certainly
understood what was required.

> The requirements for being a cosmonaut are very strict.
> They include readiness to take risks, the sense of utter
> responsibility, the ability to carry out complex tasks in
> harsh conditions, high dependability of operator's work,
> advanced intellectual abilities, and physical fortitude. …
> But the cosmonaut must also possess such qualities as curi-
> osity and the ability to break the rules. … Regulations
> work well only when everything goes as planned. … The
> ability to act in extraordinary situations is a special quality.
> In order to do that, one has to have inner freedom (even
> with respect to the Regulations), to have intuition, to have
> the ability to make non-trivial decisions and to take non-
> standard actions. In an extreme situation the very life of the
> cosmonaut depends on these qualities. Sergei Pavlovich
> Korolev understood this very well, and he captured his
> vision of professional qualifications for the cosmonaut in
> a short but capacious phrase: for cosmonauts, one must
> not select the disciplined, but the intelligent.

But for Vostok 5, trouble began soon after Bykovsky arrived at
the pad. Neither of the shortwave transmitters on the Vostok was
working; later there was a problem with the ejection hatch, and
then a control failure in the third stage. Engineers moved in

– they had just six hours to repair the faults otherwise the launch would have to be scrubbed and the Vostoks would exceed their design lifetimes. Finally, the task was accomplished but the problems were not over. In the final minute of Bykovsky's countdown a light indicated that the rocket had not severed its umbilical electrical connection to the pad. Korolev looked on the verge of panic but all around him said they should launch. In the end the rising rocket tore the cable from its socket and left it flailing on the pad.

Tereshkova and her backup Solovyeva were prepared for the second mission. All seemed to go well and Tereshkova lifted off two days later, becoming the first woman in space. The Vostoks flew closest to each other immediately after launch when they passed at a distance of about 5 km. Bykovsky later reported that he had not spotted Vostok 6, while Tereshkova thought she might have glimpsed Vostok 5. They established radio contact shortly afterwards and within three hours of the launch Moscow TV was showing live shots of Tereshkova in her capsule. But she was not feeling well: subsequent transmissions showed her tired and looking weak. She initially failed to perform one of the major goals of her mission, the manual orientation of her space-craft. Kamanin ordered Gagarin, Titov, and Nikolayev to radio new instructions. Eventually she completed the tasks, showing that if the automatic system failed she would be able to put the craft into the correct orientation for re-entry. Bykovsky reported that there had been a knock, which caused consternation on the ground. When questioned further, Bykovsky clarified what he had said: 'there had been the first space stool.' The Russian word for 'stool' (*stul*) had been mistaken for the word for 'knock' (*stuk*). It was a historic moment of sorts – the first time a human had had a bowel movement in space!

Tereshkova landed without incident, although she bruised her face. During his re-entry, as with Gagarin and Titov, Bykovsky's instrument compartment failed to separate on time from the descent capsule. It was getting to be a persistent problem. Fortunately he landed without too much worry. It was heralded as a triumph for the Soviet Union: a woman had flown in space for longer than all the six Mercury flights combined. And Bykovsky had claimed the world duration record for a single-crew spaceship: it still stands today.

Four months later, Tereshkova married Vostok 3 pilot Nikolayev in what was possibly a public relations exercise. All the top leaders of the space program attended, including Korolev, who although allowed to attend the ceremony, could not sit close to Khrushchev or Tereshkova. Even so, a few days afterwards the *New York Times* ran an article mentioning him by name as a key figure in the Soviet space effort. Tereshkova would give birth to a daughter, Elena Andrianovna (who became a doctor and was the first person to have both a mother and father who had travelled into space), in 1964. Tereshkova and Nikolayev divorced in 1982.

Even though there were plans for further female flights it took nineteen years until the second woman, Svetlana Savitskaya, flew into space – the authorities spurred by the pressure of impending American Space Shuttle flights with female astronauts. Ponomareva recalled: 'After Tereshkova's flight the commanders of the Center wanted very much to get rid of us. But the fact that we were regular officers presented an obstacle to such efforts. It was not so easy to get rid of us. Later, however, they found a way, but this first time they failed.'

William Anders, a fighter pilot for the US Navy, had tried to get into the US Air Force Flight Test School which produced astronauts for NASA, but its commandant, Chuck Yeager, famous as the first man to break the sound barrier, told him that he and the people running the school were looking for people with advanced degrees. So in 1959 he signed up for the Air Force Institute of Technology Masters degree program where he graduated with honours. He went back to Yeager who said: 'Oh, well that criteria has been changed' and that advanced degrees didn't count as much as flying time.

Frustrated, Anders was driving his Volkswagen one Friday afternoon going home from work in Albuquerque, New Mexico at the Air Force Special Weapons Center, where he was an engineer and an instructor pilot, when he heard over the radio that NASA was looking for another group of astronauts. You had to be a test pilot for the first two groups of astronauts and it didn't occur to him that they would change that. But for this group the radio announcer went down the list of things NASA wanted. He said the applicants had to be a graduate of Test Pilot School or have an advanced degree. Anders pulled over to the side, tuning it up, and then waiting for the next fifteen minute news cast where the '… or advanced degree' message was repeated. By the time he got home he had decided that he was going to put in an application.

'To my surprise,' he said later, 'I was asked to come down for the various physicals and tests several weeks later. And, to my increasing surprise, I kept surviving.' So when Yeager told him he had not got into USAF Test Pilot School he said, 'I have a better offer anyway.' Anders tells the story:

> I told him I had received a call from Deke Slayton to come to NASA. Yeager said that's not possible because we screened all the applicants and since you weren't a member of the test pilot school you didn't go forward. I said, 'Well, sir, I put in another application directly to NASA.' He was upset about that and actually put some energy into trying to get me kicked out of the NASA program. Fortunately he was not successful.

As early as March 1963, NASA had established guidelines for performing spacewalks during the forthcoming Gemini program. By January 1964, officials at Houston had completed the final details of the plan. Gemini 4, then scheduled for February 1965, would have an astronaut open the hatch and stand up for a short period. Perhaps the US would carry out the first spacewalk, or EVA (Extra Vehicular Activity).

Meanwhile, Yuri Gagarin wanted to return to space but government officials saw him as too important to risk on a spaceflight. To keep him on the ground in a high-visibility position Kamanin considered offering Gagarin the job of director of the Cosmonaut Training Centre but he was not enthusiastic for a desk job and declined several times. Later in the year, he finally gave in and became a deputy director, realising that for the time being he had no chance of returning into space. A new group of cosmonaut trainees arrived at Star City in 1963 for a year-long training program before assignment to future missions. While the first batch of twenty cosmonauts had been young Air Force pilots with little higher education, the new group of fifteen military officers all had higher degrees from a military academy or a civilian university.

The successor to the increasingly outmoded Vostok capsule was to be the Soyuz capsule designed with a voyage to the Moon in mind but by early 1964, it was clear to Korolev that it would not be ready until late 1964 or even early 1965. The Communist Party and the USSR Council of Ministers had already committed itself to the Soyuz in a joint decree on 3 December 1963, with its ultimate goal a manned flight around the Moon.

With the two-man Gemini flights being prepared, Soviet space officials were in a difficult situation. Their options were limited: none of the four projected Vostok missions in 1964 would compare favourably to a Gemini flight. They were all single-cosmonaut, none of them included a spacewalk and none of them would have the capability to change orbits. They were paying for their lack of organization, their departmental rivalry and wasteful dilution of effort, and their use of space for political statements. In this climate, an unlikely idea emerged, as audacious as it was dangerous.

Where it came from is lost. Some say that Khrushchev called Korolev and ordered him to convert the one-man Vostok spacecraft into a vehicle capable of carrying not two but three cosmonauts. According to Kamanin, Korolev was not pleased with the order:

It was the first time that I had seen Korolev in complete bewilderment. He was very distressed at the refusal to continue construction of the Vostoks and could not see how to re-equip the ship for three in such a short time. He said it was impossible to turn a single-seater ship into a three-seater in a few months.

But the account given by Khrushchev's son differs. He maintains that it was Korolev's idea, born of the desire to be first in space for as long as possible. It is certainly true that Korolev was thinking of a three-seated Vostok as early as February 1963 and he certainly had a pathological desire to be first, to beat the Americans at all costs. But wherever it came from, the decision to try to usurp Gemini proved to be one of the most disastrous decisions in the history of the Soviet space effort. Completely ignoring any natural progression of space vehicles and technology, it was a mere diversion, and for little gain. It was the very antithesis of what the Americans were doing – an incremental acquisition of abilities and technology. For the Russians, the space race had degenerated into little more than a circus act of one-upmanship. Ultimately it cost them the Moon.

The design and construction of the Apollo equipment was well under way but there was a problem – a problem of philosophy. The engineers that had come together to form NASA and those working under von Braun were traditional in their approach. Build a component, test it, attach it to the next component; test that, and so on – small incremental tests. But it was realized by some that this would not get Apollo to the Moon by the end of the decade, and would use up more hardware than the project could afford.

Enter George Mueller, who joined NASA in 1963 from the ballistic missile world. He was hired because Webb and Brainerd Holmes were not getting on. Holmes consistently ignored NASA politics and, worse, underestimated Webb's ego. He was on the front cover of *Time* magazine called the 'Czar of Apollo'. Webb was miffed and fired him. Rumours spread that NASA was an organizational mess. The various directors of NASA's centres were said not to understand their role, specifications of projects were incomplete, milestones existed but were not believed. It seemed that communication was a major problem. Webb found it difficult working with Bob Seamans. NASA was not working as a team and didn't seem to be able to pull things together. It was said that perhaps the Air Force should take over the Moon program.

Mueller, with his missile experience, had developed a different philosophy from NASA. Build it and then test it all together. It was called 'all-up' testing, and it initially worried those building Apollo. Mueller visited the Marshall Space Flight Center and von Braun to introduce his revolutionary idea. According to Mueller's plan, the very first flight would be conducted with all three live stages of the giant Saturn 5. Moreover, he said it should carry a live Apollo Command and Service Module as payload. And the entire flight should be carried through a sophisticated trajectory that would permit the Command Module to re-enter the atmosphere simulating a return from the Moon.

In retrospect it is clear that without all-up testing the first manned lunar landing could not have taken place as early as 1969. Before Mueller joined the program, it had been decided that a total of about twenty sets of Apollo spacecraft and Saturn 5 rockets would be needed. The first manned Apollo flights would be limited to low Earth orbits. Gradually NASA would inch its way closer to the Moon, and flight 17, perhaps, would be the first lunar landing. Listening to the idea, von Braun said it sounded reckless but then added that 'Mueller's reasoning was impeccable'. Mueller told everyone, 'You might as well plan for success, because you are going to have disasters anyway.'

Kennedy was having doubts. In a taped meeting at the White House he expressed concern that public support for the space program was waning. He also knew that any manned Moon landing would not be within his second term as president, assuming he was to win a second term. A few days later at a speech at a dinner in Houston he reaffirmed his belief in the program. Praising the technology emerging from the space program he said that 'those who say no in Houston, in Texas, in the United States, are on the wrong side in 1963'. Nobody who applauded him in the Coliseum Hotel that night ever thought that Kennedy would have just 15 hours more to live.

So, the mission to send a man to the Moon and return him safely to the Earth became the legacy of a martyred president, a duty of honour for a nation. Webb said, 'We have a fighting chance, but not much more than a fighting chance.'

In January 1964 the first Titan rocket was assembled on Pad 19, and fitted with the first Gemini spacecraft for an unmanned test. The following month the head of the NASA astronauts office, Deke Slayton, announced that the astronauts for the first manned flight of the Gemini project would be Alan Shepard and Tom Stafford. A few months later Shepard told

Stafford that he had a medical problem. This meant that the first Gemini crew became Gus Grissom and John Young. For a while it seemed to be the end of Shepard's career as an astronaut. He said:

I was chosen to make the first Gemini mission. Tom Stafford, who is a very bright young guy, was assigned as co-pilot, and we were already into the mission, already training for the mission. We had been in the simulators, as a matter of fact, several different times. I'm not sure whether we'd looked at the hardware in St Louis or not prior to the problem which I had. The problem I had was a disease called Ménière's; it is due to elevated fluid pressure in the inner ear. They tell me it happens in people who are Type A, hyper, driven, whatever. Unfortunately, what happens is it causes a lack of balance, dizziness, and in some cases nausea as a result of all this disorientation going on up there in the ear. It fortunately is unilateral, so it was only happening with me on the left side. But it was so obvious that NASA grounded me right away, and they assigned another crew for the first Gemini flight.

Meanwhile because of the prospective stunt involving three cosmonauts in a Vostok capsule, the Soyuz program – the real future of the Soviet space effort – was put on hold to present the false image that the Soviet Union was engaging in new leaps in space exploration. They called the new project Voskhod ('Sunrise'), hoping no one would realize that it was a strained and stretched Vostok packed with a very worried crew.

Konstantin Feoktistov, the resourceful engineer who played a critical role in the design of the Vostok, was on the Voskhod design team. Adding to the view that it might have been Korolev's idea after all, he later recalled how Korolev neutralized internal opposition:

> We argued that it would be unsafe, that it would be better to be patient and wait for the Soyuz spaceship to be built but in the end, of course, Korolev got his way. In February 1964 he outwitted us. He said that if we could build a ship based on the Vostok design which could carry three people, then one of those places would be offered to a staff engineer. Well, that was a very seductive offer and a few days later we produced some rough sketches. Our first ideas were accepted.

Feoktistov proposed getting rid of the ejection seat and spacesuits from the Vostok, thus allowing three men to cram into the spherical capsule in regular clothing. Many objected to this move but it was really a foregone conclusion, it would have been simply impossible to fit them in any other way. By the time that the draft plan was completed it was also clear that there would not be a tower-equipped launch escape system ready for the Voskhod launch but Korolev and his engineers took the risky step of moving on with the launch despite this blatant disregard for safety. It was said that it would be 'difficult' to rescue the cosmonauts up to the first 25 to 44 seconds of a launch. More accurately, if a failure occurred during that period the crew would be doomed.

Korolev's health was failing. In February 1964 he had a heart attack and spent several days in hospital. Doctors had prescribed a long holiday, which was delayed by urgent work. He was allowed to fly to Czechoslovakia for a brief holiday, the only time between 1947 and his death that he left the Soviet Union. Upon returning to Moscow he immersed himself in the Voskhod preparations. A drop test with an engineering version of the capsule was carried out in September. It was a disaster: the parachute hatch failed to open and it was smashed to pieces.

Then there was the question of which engineer was to fly on the risky mission. Feoktistov knew more about the design of the Vostok and Voskhod spacecraft than anyone else but most felt that he was not fit enough to be a cosmonaut. When Kamanin heard that Feoktistov was an option he was reported to have blurted out, 'How can you put a man into a space ship if he is suffering from ulcers, near-sightedness, deformation of the spine, gastritis, and even has missing fingers on his left hand?' The Air Force objected as well but Korolev backed his engineer and they eventually capitulated. Korolev yelled, 'The Air Force is perpetually jamming up the works! Looks like I'm going to have to train my own cosmonauts.'

The launch was set for the morning of 12 October. Korolev was more nervous and more irritable than anyone had ever seen him. The three cosmonauts arrived at the pad at 10.15 local time dressed in lightweight grey woollen trousers, shirts and light blue jackets. Korolev and Gagarin saw the three men up to the elevator before they removed their jackets and boots, donned slippers, and entered the spacecraft: Boris Yegorov first, then Feoktistov, followed by Commander Vladimir Komarov. The tension was higher than it had been for perhaps any other mission since Gagarin'. Without a viable launch escape system during the first minute of the mission, there was absolutely no way that the crew could be saved in case of booster failure. Korolev was so nervous he was shaking uncontrollably.

To his immense relief, Voskhod got into orbit without a flaw. Once again, the reaction from the West was unprecedented, prompting another speculation that the ultimate Soviet plan was to go to the Moon. Within two to three hours of the launch Feoktistov and Yegorov began to experience disorientation but despite this the short mission proceeded without much incident. When they landed there was relief all round. Because they had no room for three ejector seats a solid fuel braking rocket was added to cushion the impact of the capsule with the ground. The flight had lasted one day, seventeen minutes and three seconds and achieved nothing except propaganda. They had been lucky to get away with such a gamble.

It was later that day that Korolev and Kamanin heard of changes back in Moscow. News had come in that there would be a special meeting of the Central Committee the same evening. Within hours Khrushchev was no longer in power and had been replaced in his two posts by Alexei Kosygin and Leonid Brezhnev. Kamanin was instructed to alter the cosmonauts' speeches. Instead of saluting Khrushchev, they would salute Brezhnev and Kosygin.

It was clear to Korolev that the second Voskhod mission should include a spacewalk – NASA's stated plans to carry out an EVA during the Gemini program once again compelled him to try to beat the Americans. But how could a Vostok capsule be modified so that a spacewalk could take place?

Soviet engineers could not consider the Gemini approach of depressurizing the entire spaceship during an EVA because their life-support systems were not good enough and the instruments in the Vostok capsule were not designed to operate in a vacuum. Instead, they drew up a plan for an airlock, made of rubber, to be unfolded on the outside of the spacecraft. Both cosmonauts would wear pressure suits throughout the flight. Before the spacewalk the pilot would crawl into the airlock, shut the hatch behind him, evacuate it, open an outer hatch, and then step out into space. A 5-m cord would connect the cosmonaut to the ship during the EVA. The maximum time in space would be limited to ten to fifteen minutes.

Kamanin was thinking of having two women fly in a future Voskhod spacecraft, with one of them carrying out a spacewalk. He mentioned the idea to Korolev, who would have none of it, but Kamanin had ensured he had support from higher up in the program. After Tereshkova's flight, the other four female cosmonauts had been consigned to support roles, but Kamanin's new idea brought them back. In April 1965, Ponomareva and Solovyeva began training for a spacewalk during Voskhod 5. Solovyeva would be the first woman to walk in space. But the planning did not proceed well. Plant No. 918, which made the spacesuits, refused to take on the job of designing completely new spacesuits for women.

By this point, the best candidates for the primary crew of the Voskhod 2 spacewalk mission were Belyayev and Leonov. The now-39-year-old Pavel Belyayev had been the oldest candidate from the 'Gagarin group' of 1960. He had graduated from the Yeisk Higher Air Force School in 1945 and flew combat missions against the Japanese during the final days of World War Two. Later, in 1959, he graduated from the famous Red Banner Air Force Academy, and thus he was one of only two cosmonauts in the 1960 class who had a higher education. Belyayev might have flown into space earlier had it not been for a severe ankle injury sustained in August 1961 during a parachute jump, which left him out of the running for a whole year. Thirty-year-old Aleksei Leonov was born in Siberia, and graduated from the Chuguyev Higher Air Force School in the Ukraine in 1957 before serving as a jet pilot in East Germany.

Chief Designer Severin recalled: 'the Americans planned to do their EVA in three months and had announced it before-hand. So we felt very rushed. We were hurrying and were nervous.'

The first Voskhod 2 test spacecraft was launched into orbit successfully on 22 February 1965. As it was designated under the catch-all Kosmos classification the media did not realize its true function. The fully equipped spacecraft was to simulate all the necessary airlock operations. Meanwhile, the ground tests for these aspects of the mission were beset by failures. Severin recalls:

> The situation was really grave. Almost the entire testing program had been disrupted. Only part of it was com-pleted in the unmanned flight. There was even talk of postponing the flight until better results were obtained on the ground. The competition with Gemini reached such

a state that Soviet security personnel arrived at Baikonur. It's possible that the KGB thought that all of our accidents were the result of sabotage. They imposed strict monitoring, which made us very nervous.

Korolev was beset by poor health leading up to the launch. At one point, he had had to spend some time under medical attention because of a pulmonary inflammation. But nothing deterred him from his work, and looking tired and gaunt, he showed up for the launch on the morning of 18 March. It was a cold and snowy day at Baikonur. He said to Leonov, 'May you have a fair solar wind.'

Voskhod 2 lifted off successfully and the two cosmonauts began preparations for the EVA as soon as they reached orbit. First, Belyayev expanded the rubber airlock to its full length. Then Leonov, aided by Belyayev, strapped on his life-support pack. Once the pressure between the airlock and the ship was equalized, Belyayev opened the inner hatch allowing Leonov to crawl headfirst into the airlock and hook himself up to the tether. Then Belyayev shut the inner hatch and depressurized the airlock. Leonov emerged, becoming the first human to walk in space. The Sun almost blinded him. His first words were: 'I can see the Caucasus.' But after twelve minutes in open space Leonov found himself in a perilous situation:

> Near the end of my walk I realized that my feet had pulled out of my shoes and my hands had pulled away from my gloves. My entire suit stretched so much that my hands

and feet appeared to shrink. I was unable to control them. I couldn't get back in straightaway. My space suit had ballooned out and the pressure was quite considerable. I was tired and couldn't go in feet first as I had been taught to do.

Leonov decreased the pressure in his suit hoping that it would make it more flexible.

Then I felt freer and I could move about more easily. Then I pushed myself into the airlock head first, with my arms holding the rails. I had to turn myself upside down in the airlock in order to enter the ship feet first and this was very difficult.

His pulse raced to 143 beats per minute, his breathing was twice normal levels, and his body temperature was up to 38°C. He was drenched in sweat and exhausted. Finally, the outer hatch was closed, giving a total depressurized time of 23 minutes and 41 seconds. They cast off airlock and settled down to a one-day mission.

But there was another problem. The hatch on the ship had not been shut properly and was leaking air, which was being compensated by the life-support system. The result was that the capsule was becoming rich in oxygen, which increased the possibility of a fire. A tiny spark could set off an explosion. They lowered the oxygen during the rest of the mission, bringing it down to manageable levels before re-entry. It would not be the last time that an oxygen-rich atmosphere inside a capsule would pose a risk.

The problems kept coming. By the thirteenth orbit, pressure in the fuel tanks had dropped dramatically, bringing with it the possibility of the complete depressurization of the spacecraft. Fortunately it stabilized. When the re-entry burn came around on the seventeenth orbit, Belyayev calmly reported, 'Negative automatic retrofire.' Korolev immediately told Belyayev to use the manual system. The numbered code to unlock the attitude controls was found and was handed to Gagarin, who transmitted the information to Belyayev.

The exercise of orienting the spacecraft became an ordeal, if not a fiasco. They had to use an optical sighting device but both men were clad in bulky spacesuits. In the cramped space, Belyayev, optical device in hand, had to lie horizontally across both seats while Leonov tucked himself under his seat. At the same time, Leonov held Belyayev in place so as to keep him in front of the porthole so he could use both his hands to orient the ship with respect to the Earth's day–night divide, or terminator, as it is called, using the hand controls. After this was done they quickly returned to their seats to re-establish the ship's centre of gravity before firing the retrorocket. The 46 seconds it took to get back resulted in a serious overshoot of the original landing zone. As with several previous missions, the instrument compartment failed to separate from the descent capsule and the two modules remained connected by the flailing steel straps. It resulted in a steeper than usual descent and more G force, bursting blood vessels in both men's eyes.

They landed in dense woodland, far from any settlement. One of the pilots in a search helicopter reported: 'On the forest road between the villages of Sorokouaya and Shchuchino. About 30 km southwest of the town of Berezniki, I see the red parachute and the two cosmonauts. There is deep snow all around.' The denseness of the trees made it impossible for helicopters to land. Warm clothing was dropped from one, while another landed 5 km away. When Leonov asked how soon rescuers would pick them up, Belyayev joked, 'Maybe in three months, they'll pick us up with dog sleds.' They spent an uncomfortable night in the woods.

Incredibly, ignoring the obvious risk they had taken, Korolev raised a toast to the future: 'Friends! Before us is the Moon. Let us all work together with the great goal of conquering the Moon.' It was not to be.

The first manned Gemini mission, Gemini 3, with Grissom – becoming the first person to make two spaceflights – and Young on board was launched on 23 March 1965. It was a brief flight of just three orbits lasting a total of just under five hours. During that time they changed orbits, at one time achieving an orbit that had a low point of just 85 km. Their flight was successful but after splashdown the build-up of heat within the capsule was too much for the astronauts so they abandoned Gemini 3 in their spacesuits, and they walked along the traditional red carpet on the recovery ship in their underwear! After splashdown Grissom was also seasick. 'Gemini may be a good spacecraft but she's a lousy ship,' he said.

Commander of Gemini 4, the mission that was to undertake the first US spacewalk, was Jim McDivitt. He later commented that by the time the US started Gemini, the space race was over. He talked about the importance of the Gemini program:

In Mercury, you couldn't manoeuvre. You could change its attitude but you couldn't change its flight path. Gemini you could. So, now you had to have the guy in the spacecraft working with the guy on the ground to know what was going on and where they were going, where they were, and […] what was going to happen. So, that worked out pretty well. As a matter of fact, I think if it hadn't been for Gemini, flying Apollo would've been nigh on impossible.

McDivitt tells a funny story about getting picked to fly on Gemini 4:

I was called in and told I was going to command it, and then some time later it was announced at an astronaut pilots' meeting and then finally they were going to make the public announcement. And so, I thought I'd tell my kids about it. So, one Saturday morning we were sitting having breakfast at a long table we had. And so, we finally got to this dramatic moment and I said, 'Kids, I'm going to tell you something really important.' And, let's see, this was in about '64 or so. I think they were, like, eight and seven and five or so. And so, I tell them that, you know, 'You know that dad's an astronaut and the astronauts fly in space. I just want to let you know that I'm going to fly in space soon.' And my older boy, Mike, who was probably seven or eight, says, 'Oh yeah, dad, I heard that at school.' And then my daughter Ann said, 'Oh yeah, dad, I heard that at school, too.' And my son Patrick said, 'Dad, there's a fly in the milk bottle.'

The USA's first spacewalk was carried out by Ed White, of whom McDivitt said:

> My relationship with Ed couldn't have been better. He was the best friend I ever had. We lived, like I said, a block and a half or so apart. He was getting a Master's degree in aeronautical engineering, but he didn't have an aeronautical engineering undergraduate degree. So, we took a lot of classes together. We started flying together. I remember when the Air Force had its pre-astronaut, pre-NASA astronaut selection – I walked in the room in the Pentagon and Ed was already there and he says, 'I knew you'd be there!' And I said, 'I knew you'd be here, too!' Unfortunately as regards our EVA we were beaten by the Russians by, what? A couple of weeks I guess. They were quiet up until a few days before the flight. I think originally it was to score the first!

Gemini 4 was dispatched to space on 3 June – the USA's first multi-day mission – and once in orbit they turned their attention to the spacewalk. McDivitt recalls:

> When we got around to doing the EVA, Ed went to open up the hatch, but it wouldn't open. I said, 'Oh my God,' you know, 'it's not opening!' And so, we chatted about that for a minute or two. And I said, 'Well, I think I can get it closed if it won't close.' But I wasn't too sure about it. I thought I could. But remember, then I would be pressurized. I wouldn't be in my sports clothes, leaning over

the top of the thing with a screwdriver as I had been in training. I'd be there pressurized. In the dark. So anyway, we elected to go ahead and open it up.

White was outside for 21 minutes and had to be told to come back in by the Capcom Gus Grissom. Gemini 4 was headed for the Earth's shadow. 'This is the saddest moment of my life,' replied White.

McDivitt continues the story:

That was one of the reasons I was kind of anxious to have him get back inside the spacecraft, because I'd like to do this in the daylight, not in the dark. But by the time he got back in, it was dark. So, when we went to close the hatch, it wouldn't close. It wouldn't lock. And so, in the dark I was trying to fiddle around over on the side where I couldn't see anything, trying to get my glove down in this little slot to push the gears together. And finally, we got that done and got it latched.

For Gene Kranz, Gemini 4 was one of the most exciting of the Gemini missions. It was his first as Flight Director:

We wanted to be the first to have an extravehicular operation; put a man out in space, free from the spacecraft. I got tagged to work with the team in building that EVA plan. And we were very imaginative; we called it Plan X. We'd finish our work here during the day; we'd go home, we'd eat, and then all the Plan X people would come back in and we'd work generally from about six or seven in the evening until one or two in the morning, building the equipment, validating it in the altitude chamber, developing mission rules, etc.

I was one of those who believed in the space race. I wanted to beat the Russians. I didn't like Russians. I'd seen their airplanes over in Korea; I'd seen them over the Formosa Straits. And, to put it bluntly, it was a battle for the minds and the hearts of the free world. So, space was not just something romantic to me. It was the battleground with the Soviet Union at that time. I really wanted to set this first space record. Well, the unfortunate thing is, the Russians had already accomplished extravehicular operations. But, the neat thing about this was, we now knew when they had this enormous lead on us to begin with that this lead was now down to mere months. When they were doing their EVA, we were within striking distance of that EVA.

In June, following their spectacular Gemini 4 flight, James McDivitt and Edward White flew up to Washington from Houston with their wives and children. The helicopter bearing them had no sooner settled on the White House lawn than Lady Bird Johnson said she wanted them all to spend the night; babysitters would be provided. The two astronauts heard the President call them 'Christopher Columbuses of the 20th century', and pronounced that the United States had now caught up with the Russians. They lunched with Vice President Hubert Humphrey and Congressional leaders, and in the evening they went to the State Department for a reception.

Before a packed assemblage of foreign diplomats they showed a 20-minute movie of their flight, which included the first American spacewalk. In strode Lyndon B. Johnson, who told McDivitt and White, 'I want you to join our delegation in Paris.' Furthermore, the President wanted them to go now, as soon as they and their wives could pack. He was seething because the Russians had humbled the Americans at the Paris Air Show, where Yuri Gagarin was standing by his spacecraft, shaking hands with everybody and passing out Vostok pins. The French press noted that the lacklustre American pavilion was shunned by the crowds.

Patricia McDivitt and Patricia White wailed in unison, 'But we have nothing to wear!' Never mind, said LBJ, Lady Bird and Lynda and Luci (his daughters) have plenty of clothes. The ladies retired to the White House bedrooms, and the two Patricias were duly outfitted. Long after midnight the presidential plane took off with them on board as well as Vice President Humphrey, James Webb, and Charles Mathews, the Gemini program manager. They made it only in time for the last day and a half of the eleven-day show, but they gave the Russians some real competition. Wherever they appeared, the American *jumeaux de space* were followed by crowds. 'A partial recovery for the United States' was the Paris newspapers' verdict.

Two months later Gemini 5 was launched with Gordon Cooper and Pete Conrad on board. Cooper says:

> Our Gemini 5 flight. We worked long and hard at it, and we couldn't do any EVA or do the other things because we were so loaded. We were absolutely crammed with

equipment of all kinds they wanted us to have. We had the first fuel cell. We had the first radar. We had the first all up computer. These were all things that needed to be tested and proven. So we named it 'Eight Days or Bust'.

The Gemini flights before that didn't have any name or any patches; and Pete and I decided that since everybody in the military, every pilot in the military, has an organizational patch, and they take great pride in their patches, we decided we were going to have a patch. So we designed and had some made, and had them sewn on our flight suits; and two nights before the launch, we were invited to fly back to Houston and have dinner with Jim Webb (NASA Administrator) at Bob Gilruth's house. We decided this would be an opportunity to confess our sins, so we told him about the patch, with which he went almost into hysterics! And he and Pete almost had fisticuffs. So we got this all broken up; and he finally said, 'Okay. Tomorrow I want you to send somebody up to Washington with a copy of this patch so I can see what it is, and I'll tell you whether you can fly it or not.' So, the following day, he took a look at the patch and then wired us back down an approval. He said, 'From now on, it'll be called a "Cooper patch". I will approve it on one condition.' He said, 'I don't like the "Eight Days or Bust" cause if you only do seven days, then the public's going to say, "You busted."' So he said, 'If you'll tape over the "Eight Days".'

At the start of Gemini 5 in August 1965 the oxygen pressure dropped to practically zero. According to the mission rules

The 'Mercury 7' astronauts photographed on 9 April 1960.
Back row: Alan Shepard, Gus Grissom, Gordon Cooper.
Front row: Wally Schirra, Deke Slayton, John Glenn, Scott Carpenter.
This was the only time they would appear together in pressure suits.
NASA

(*above*) Wernher von Braun a the Redstone Arsenal in Alab in the mid-1950s when he wa head of the US Army's rocket development team.

Sergei Pavlovich Korolev (1907–1966) the Soviet 'Chief Designer'.

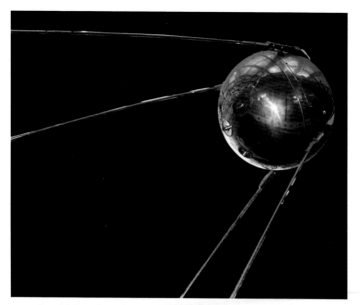

Sputnik 1 was launched on 4 October 1957. It was simple in design
– basically three batteries and a radio transmitter. Its 'beep, beep, beep' changed history.

or Yuri
:eyevich
;arin in pressure
and helmet on
round 12 April
1, when he
me the first
in space.

US President John F. Kennedy presents astronaut Alan Shepard with the NASA Distinguished Service Award at the White House on 8 May 1961.

Gemini 6-A (foreground) and Gemini 7 make the first rendezvous in orbit between two manned spacecraft, 15 December 1965.

Jim Lovell, capsule communicator (Capcom) for the Gemini 8 mission, 16 March 1966.
Neil Armstrong and David Scott had to abort the mission and return to Earth early.

Neil Armstrong and David Scott await their recovery craft the USS *Leonard F. Mason*
after splashing down, 16 March 1966.

The crew of Apollo 1: Virgil I. 'Gus' Grissom, Edward H. White II and Roger B. Chaffee. All were killed in a pad fire on 27 January 1967.

NASA

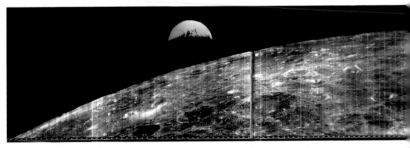

The first photograph of Earthrise over the Moon, taken by Lunar Orbiter 1 on 23 August 1966

NASA

One of the most famous pictures ever taken. Earthrise as seen by the crew of Apollo 8.
Picture taken by Bill Anders, 24 December 1968.

NASA

The Apollo 9 Lunar Module being tested in Earth orbit in lunar landing configuration, 7 March 19

Gene Kranz was Mission Controller for several Apollo missions.
Photo taken 16 April 1972, during Apollo 16.

NASA's official Apollo 11 crew portrait. From left to right: Neil Armstrong (Commander), Michael Collins (Command Module Pilot) and Edwin 'Buzz' Aldrin Jr (Lunar Module Pilot).

Neil Armstrong and Buzz Aldrin practise landing in a Lunar Module simulator, July 1969.

Apollo 11 launch,
16 July 1969.
NASA

(below) Pat Collins, w
of Michael Collins,
with their children
Mike, 6; Ann, 7; and
Kathleen, 10, after th
successful launch of
Apollo 11.
NASA

Charlie Duke, Capcom for the landing. To his left are Jim Lovell and Fred Haise.

Flags and cigars in Mission Control as Apollo lands on the Moon.

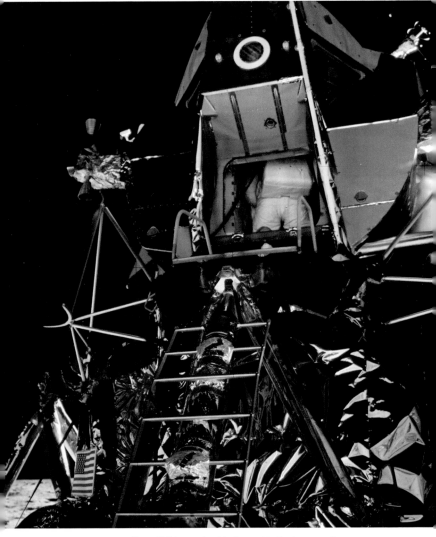

Buzz Aldrin starting his descent to the lunar surface,
as photographed by Neil Armstrong.

Buzz Aldrin on the lunar surface.
Neil Armstrong can be seen in his visor.

NASA

Neil Armstrong back in the Lunar Module after the first moonwalk.
NASA

The ascent stage of the Lunar Module, 'Eagle',
returns to the Command Module, 'Columbia', 21 July 1969.
NASA

The Apollo 11 crew in the mobile quarantine facility in which they spent two and a half days
route from their recovery ship USS *Hornet* to the Lunar Receiving Laboratory in Houston, Texas.
NASA

Apollo 13's damaged service module after being jettisoned, 17 April 1970.
NASA

The last man to walk on the Moon, Eugene A. Cernan rests in the Lunar Module after the third and final moonwalk undertaken with his crewmate Harrison Schmidt, 13 December 1972.
NASA

the correct procedure was to switch everything off. Says Cooper:

I had to go into total power down. So we powered everything down, brought everything down to low, low electrical power; and, of course, it happened again when we were out of radio range. So, as we came whistling over the horizon into communication, Houston realized we were all powered down and they really panicked for a moment; and it looked like we were going to have to re-enter another orbit later. But fortunately, and this is a story a lot of people don't know, when Pete and I were going through the altitude chamber with Gemini 5, we had to go through and do these various tests; and the tests finished on a Friday; the spacecraft was due to be shipped Saturday morning to the Cape from St Louis in order to stay on the time schedule. But one of the things we wanted to do was, we wanted to run some tests in the altitude chamber by decreasing both oxygen and hydrogen pressure, drastically, to see if the fuel cell would continue to run at altitude. NASA said, 'No, we can't afford the time […] and if we do it over the weekend, it would cost us triple time of overtime, so we're not going to do it.' So Pete and I went to Jim McDonnell, head of McDonnell Aircraft Corporation and told him the story on it, and he said, 'I'll pay for it. Let's do it.' So we spent the weekend in the altitude chamber at his cost doing the tests.

The knowledge they gained enabled them to stay in orbit longer.

Finally, the United States had taken the space endurance record. Korolev extended the planned Voskhod 3's duration from ten to fifteen days and then to twenty days. Then at a meeting of the Military-Industrial Commission on 16 December 1965, the Soviet government added one more condition to the Voskhod program: that Korolev launch two Voskhods in time for the 23rd Congress of the Communist Party in March 1966 as a salute to the Party. It was an unrealistic deadline. Things were falling apart. In the event, the Russians did not launch another manned mission until April 1967 – and then it was a disaster. In the meantime the US launched six manned flights, completing the Gemini project. The next of these were Gemini 6 and 7 – a double mission. Frank Borman flew Gemini 7 with Jim Lovell on the record fourteen-day mission. According to Borman:

Gemini 7 was looked upon among the astronaut group as, you know, not much of a pilot's mission. Just sort of a medical experiment mission, which it was. Jim Lovell was a wonderful guy to spend 14 days with in a very small place. We had a lot of interesting things. You know, some of the doctors said, 'Oh well, in order to do that you're going to have to simulate it on Earth and see if you can stay in one G for 14 days.' And I, you know, 'They're out of their mind. Fourteen days sitting in a straight-up ejection seat on Earth? You're crazy!'

We'd been up there for 11 or 12 days (I don't remember how long). And we were tired, and the systems on the spacecraft were failing. We were running out of fuel, and it was a real high point to see this bright light (it looked like a star) come up, and then eventually we could see it was a Gemini vehicle – Gemini 6. And we found that we could – we had very limited fuel – but we found that the autopilot for the controls were perfect. You could fly formation with no problem [...] about the only thing that I really felt after two weeks like that were our leg muscles were shot. And it took about three or four days; and I guess you could feel it for a week or so afterwards. But it wasn't any big deal.

In fact, Wally Schirra and Tom Stafford nearly didn't get Gemini 6 into orbit. The plan was for Gemini 6 to launch before Borman's flight and dock with an Agena target vehicle already in space, but the Agena didn't get into space. So a homing device was put on Gemini 7 for Gemini 6 to use.

Things were not going well in the USSR. Their unmanned lunar reconnaissance probes were failing. Between January 1963 and December 1965, there had been eleven consecutive failures, a record that had dampened the spirits of even the most optimistic Soviet engineers. Kamanin wrote in his diary: 'Korolev was more distressed by the setback than anyone. He looked dejected and appeared to have aged ten years.'

In fact, Korolev's health had been deteriorating seriously throughout 1965. In August he had complained about not feeling well because of low blood pressure, and in September, he had severe headaches. He also had a progressive hearing loss and a serious heart condition. He wrote to his wife: 'I am holding myself together using all the strength at my command ... I can't continue to work like this, you understand. I'm not going to continue working like this. I'm leaving!' He knew the Voskhods would be unable to achieve what Gemini had.

Worried by the stagnation, a group of experienced cosmonauts, along with Kamanin, wrote a letter to Brezhnev on 22 October 1965. They emphasized the gridlock in the space program because of its complicated management system, the undue focus on automated systems over piloted ones, and the poor funding. Gagarin personally handed the letter to Brezhnev's aides, but three months later, they were still waiting for an acknowledgement.

Plugs Out

On 10 August 1966 came the first of the next set of US Moon missions. Lunar Orbiter 1 was launched and it entered orbit of the Moon four days later, returning its first pictures four days after that. Five Lunar Orbiter missions were launched in 1966 and 1967 to map the surface before the Apollo landings. All were successful, and 99 per cent of the Moon was photographed with a resolution of 60 metres or better, ten times what could be achieved by ground-based telescopes. The first three missions were in equatorial orbits focusing on twenty potential landing sites. The fourth and fifth were devoted to broader scientific objectives and were in high-altitude polar orbits. Lunar Orbiter 4 photographed the entire near side and 95 per cent of the far side, and Lunar Orbiter 5 completed the far side coverage.

On 23 August Lunar Orbiter 1 took a picture that caused a sensation, the Earth hanging over the lunar limb. There was our planet lonely and fragile floating in the great cosmic dark. Every person who had ever lived, all known life was confined to a thin skin on that cloud-covered ball we could now all see. In the vastness of space it was an oasis. There are some who maintain that today's strong environmental movement stems from that picture.

The Soviet Luna 11 entered lunar orbit on 28 August 1966. Luna 12 achieved lunar orbit on 25 October 1966. These spacecraft were equipped with a television system that obtained and transmitted photographs of the lunar surface but they were nowhere near as good as those from the Lunar Orbiters. The technological balance between the superpowers was shifting.

In November 1966, five days into an eight-day mission, Lunar Orbiter 2 was being prepared for its next mapping sequence and turned to take a picture looking at the ejecta ray systems around the 100-km-wide Copernicus crater – scientists were unsure if rayed areas were smoother or rougher than the surface they overlaid. The oblique image of wide Copernicus was immediately called the 'picture of the century'. In the foreground were the 300-m-high mountains of the crater's central peak system. From what is understood about the dynamics of the impact that caused this crater about a billion years ago (young in terms of the Moon's age), the rocks in those central peaks have been raised from far below the Moon's surface. Analysing them would reveal much about the history and the origin of the Moon. No wonder many astronauts looked at the picture and pointed at those peaks and said 'Land there.' Beyond the complex peak system are the ramparts of the crater, and beyond that the Carpathian Mountains.

Surveyor 3 landed on 20 April, touching down near the crater Lansberg in the Ocean of Storms. It bounced several times before settling in a 200-m-wide crater. It had a remote-controlled arm fitted with a scoop; it found that the surface was soft with hard rock at a depth of 15 cm. The next mission occurred on 14 July 1967, when Surveyor 4 was aimed for a landing in the Central Bay. It was the second attempt to land a Surveyor in the centre of the Moon and it failed. Surveyor 5, the first in the Surveyor series to carry a scientific instrument for analysis of the Moon's surface, successfully landed on 10 September 1967 in the Sea of Tranquillity not far from where in less than two years men would walk. During the two-week-long lunar day it operated perfectly and transmitted a total of 18,006 television pictures, greater than the combined total from two previously successful Surveyors. The analysis indicated that the lunar surface near the spacecraft was composed of a basaltic rock similar to that found in Greenland.

Surveyor also achieved a minor first. Mission 6 soft-landed on 9 November in Sinus Medii – the Central Bay in the centre of the Moon as seen from Earth – and within a week had sent back nearly 10,000 images as well as data on the composition of the lunar soil. Then its thrusters were fired once again, just two and a half seconds, to lift it four metres and move it a few metres to one side. It was the first ever lift-off from another world. It landed upright and started taking pictures again 35 minutes later. Its stereo camera looked back at where it had rested and showed the imprint of the footpads at the original landing site. One small step for a machine.

With the Surveyor mission goals complete, Surveyor 7 was, like the last of the Rangers, given to the scientists, who chose to send it to one of the Moon's most spectacular features, the crater with the magnificent ray system that dominates the face of the full Moon. It was to land on the rim of Tycho in the Southern Highlands. On 10 January 1968 it touched down some 30 km from its rim. Scoops of the soil showed it to be like feldspar. This type of rock suggested that the entire surface had once been molten, before the heavier elements sank out of a magma ocean, with the lighter elements floating to the surface to create a fine crystalline crust. It was a vital piece of evidence concerning the origin of the Moon.

1966 was a game-changing year for the Soviet space effort as well, although in a very different way. In December 1965, Korolev had undergone a series of medical tests in Moscow, which indicated he needed a minor operation to remove a bleeding polyp in his intestine. He spent his last day before the operation – 4 January 1966 – at his office, staying late before being admitted to the Kremlin hospital the following day. He had already invited people to celebrate his 59th birthday at a party on 14 January.

Dr Boris Petrovsky, the USSR Minister of Health, removed a small polyp from Korolev's gastrointestinal tract, causing excessive bleeding. Petrovsky was an accomplished surgeon but it seems he wasn't prepared for the operation. There were complications. Korolev had not told them that his jaw had been broken in prison (during torture) in 1938, which made it difficult for him to open his mouth wide. His unusually short neck compounded the problem, preventing the use of an intubation tube into his lungs. Instead, the surgeons performed a tracheotomy, inserting a tube in his neck. Korolev bled profusely during the operation and then Petrovsky found what he later described as an 'immovable malignant tumour which had grown into the rectum and the pelvic wall'. The size of the tumour, larger than a person's fist, was a shock to those in the operating room. Dr Vishnevsky, a cancer specialist, was called in and the two surgeons completed the operation, but half an hour later, Korolev's heart stopped and they could not revive him.

His death shocked the Soviet space effort but Korolev's arch-enemy Glushko, who had worked in a separate design bureau for years, coming up with alternatives to Korolev's plans, was unperturbed. He was conducting a meeting when his Kremlin phone line rang. He heard the news, hung up, and turned to his audience and said, 'Sergei Pavlovich is no longer with us.' He paused for a second and continued, 'Now where did we leave off?'

Korolev was given a state funeral on 18 January. The urn with his ashes was carried from the House of Unions by Gagarin and others. In Red Square Brezhnev and other Soviet leaders placed it in the Kremlin Wall.

Mishin – who had been with Korolev since they were both scavenging for V-2 parts in 1945 – was clearly the most likely choice as a successor, having been groomed for the role by Korolev for almost a decade. His first job as was to assess the state of the Voskhod program. At the time of Korolev's death, there were plans for three to four Voskhods and five Soyuz missions in 1966. The first one, Voskhod 3, was to be a long-duration mission with cosmonauts Boris Volynov and Georgi Shonin. The spectacular success of the fourteen-day Gemini 7 flight in December 1965 had given the Soviet mission even more of an impetus. The subsequent Voskhod 4 would be a scientific flight, including artificial gravity experiments with test pilot Georgi Beregovoi and scientist Georgi Katys, while Voskhod 5 would be a military mission that included cosmonaut Vladimir Shatalov.

A Voskhod test spacecraft with two dogs was launched in February. The flight lasted nearly 22 days and when physicians examined the dogs upon their return they saw what a dreadful condition they were in: wasting muscles, dehydration, calcium loss, and confusion in readjusting to walking. Their motor systems did not return to normal until eight to ten days later. While Mishin and his colleagues were assessing what had happened, the United Press International Agency reported that the Soviet Union would launch a multi-crewed spacecraft before the end of March 1966, in time for the 23rd Congress of the Communist Party. This was the planned Voskhod 3 mission, by which they intended to regain the duration record. It was to be an outstanding publicity victory for the Soviet space program, but long-duration ground tests of the life-support system did not go well. After fourteen days, the Institute for Biomedical Problems had to abandon them because of a worsening of the cabin atmosphere. Parachute failures during recovery tests were common. Four cosmonauts were in training for the flight but as the problems accumulated it became increasingly clear that there might never be a Voskhod 3 mission. Soon it was cancelled.

All the momentum seemed to be with the Americans. But where their previous manned flight, Gemini 7, had been a routine mission, Gemini 8 certainly would not be. It was to be NASA's first serious space emergency.

Neil Armstrong had become a civilian test pilot, flying the advanced X-15 rocket-plane out of Edwards Air Force Base in California. Apollo excited him and he applied to be an astronaut. Deke Slayton called him in September 1962 and he became a member of the so-called 'New Nine'.

Dave Scott came in the following year's selection, and he and Armstrong were launched in Gemini 8 on 16 March 1966, following an Agena target vehicle into space. Just a few hours after lift-off Armstrong manoeuvred the spacecraft to make the first docking in space. Then the trouble began. Without warning they went into a dangerous spin. Armstrong said later:

We first suspected that the Agena was the culprit. We had shut our own control system off, and we were on the dark side of the Earth, so we really didn't have any outside reference, or very good reference. I didn't actually notice when it started to deviate from the planned attitude. Dave first noticed it. Neither of us thought that Gemini might be the culprit, because you could easily hear the Gemini thrusters whenever they fired. They were out right in the nose, in the back. Every time one fired, it was just like a popgun, 'crack, crack, crack, crack.' And we weren't hearing anything, so we didn't think it was our spacecraft. Dave had the control panel for the Agena. That allowed Dave to send signals to the Agena control system. He was trying everything he knew, without success.

When the rates became quite violent, I concluded that we couldn't continue, that we had to separate from the

Agena. I was afraid we might lose consciousness, because our spin rate had gotten pretty high, and I wanted to make sure that we got away before that happened. Of course, once we separated and found out we couldn't … regain control in a normal manner, we recognized that it was a failure in our craft, not in the Agena. The reason we didn't hear it is, you only hear the thruster when it fires; you don't hear it when it's running steadily. I didn't know that at the time, but I figured it out.

When Armstrong and Scott undocked from the Agena 'all hell really broke loose', according to the mission's Flight Director John Hodge.

Jim Lovell could have been one of the Mercury 7 but didn't make it for a trivial medical reason that was overruled when he joined the astronaut corps at the same time as Armstrong. He was the capsule communicator at the time:

It turned out finally that it was one of their thrusters that was firing. And so, they managed – and they were very cool about this whole thing – they managed to pull the circuit breaker and get the thruster offline. And then using their – the re-entry thrusters, which were normally only used for re-entry, they were able to slow the vehicle down; and, of course, they had to come back early. When they came back into radio contact, well, they said, 'Hey,' I guess they used to use the old phrase long before [Apollo] 13, they – 'we got a problem.' If they didn't correct the problem themselves, they would've been in deep trouble.

Sometimes, like on Gemini 8, we were just lucky. And luck's got no business in spaceflight.

'With our vision beginning to blur,' wrote Scott, 'locating the right switch was not simple.' Scott was amazed at Armstrong's skill as he reached for the toggle and grappled with the spacecraft's hand controller at the same time.

Television stations began interrupting their programs – *Batman* and, ironically, *Lost in Space* – to provide live coverage. The retrorockets fired above South Africa and the spacecraft re-entered over the Himalayas. Scott could see nothing through his window. Ten hours and 41 minutes after leaving Cape Kennedy, Gemini 8 splashed down. 'When Mission Control told us about three-foot waves,' Scott recalled, 'they forgot to mention the 20-foot swells!'

Robert Seamans, NASA's Deputy Administrator, was at a dinner when he was told about the problem. The cool-headedness of Neil Armstrong and David Scott in a life-threatening situation did not go unnoticed. Both of them would walk upon the Moon. Afterwards Armstrong described it as a 'non-trivial situation'.

The incident provided an insight into the complicated role of the Flight Director. John Hodge had followed the mission rules and played it by the book. But he didn't consult his boss, the Gemini program manager or the director of the space centre, who were all miffed that they hadn't been consulted and who took the view that a couple of hours' wait would not have endangered the crew. John Hodge never served as Flight Director in Mission Control again.

Gemini 9's crew was Tom Stafford and Gene Cernan (second and third astronaut intake respectively). They were supposed to dock their Gemini with an Agena and perform a lengthy spacewalk but the Agena's protective metal shroud had failed to come away, giving it what was described as an 'Angry Alligator' look. As Tom Stafford was suiting-up for the flight, Deke Slayton, head of astronaut assignments, appeared and told the suit technician to leave the room. Slayton said, 'I need to talk to you, Tom.' Stafford takes up the story:

He [Slayton] said, 'Look, this is the first time we've got this long EVA, this rocket pack, and NASA management's decided that in case Cernan dies out there, you've got to bring him back, because we just can't afford to have a dead astronaut floating around in space.' […] He left. Gene says, 'Hey, Tom, Dick was in there talking to you quite a while. What did he say?' I said, 'He said he just hoped we'd have a good flight.'

So we got all ready to go and we launched, and it lifted off in June. I remember coming up to it, and you could see the constellation Antares. There was a full Moon out. We got up close, I could see this weird thing. I came right up close to it, and it just broke out in sunrise, and here was the shroud, like that. I call it 'The Angry Alligator'. Then came time for Cernan to go EVA, and they wanted him to go out and cut loose the shroud, to cut it loose. I looked at it. I could see those sharp edges. We had never practised that. I knew that they had those 300-pound springs there, didn't know what else. So I vetoed it right there. I said, 'No way.'

'Cernan goes out, and the first thing he does is place the rear-view mirror on the docking bar. He's huffing and puffing. He's torqueing the hell out of this spacecraft, and I'm pulsing it back to be sure none of the thrusters fire on him. He goes out in front and does a few little manoeuvres and he's having a very difficult time. […] Then he says, 'Tom, my back's killing me. It's burning up. It's really killing me.' I says, 'What?' He says, 'My back.' I could look in the rear-view mirror, and I could see the Sun. Of course, you never look directly at it. I said, 'Do you want to get out of the spacecraft?' He said, 'No. Keep going, but my back is killing me. It's burning up.'

So he finally, just before sunset, gets turned around into the seat. We had two lights back there. One of them burned out for some reason, during the vibration of the launch anyway. We had one light. And then a couple of minutes after sunset, he was strapping himself in. […] He fogged over. Whop! He could not see. It was just like that, fog. So we did defog on the visor, and he had overpowered that little water evaporator so much, it was unbelievable.

As if the conditions weren't bad enough for Cernan, they were having problems with the radio. Stafford continues:

I think we went south of Hawaii, then, before we hit the West Coast of the U.S. We went a long time. It was night time. We saw the Southern Cross go by. What a hell of a lonely place this is. Here you're 165 miles up, you know, flying, pressurized. Your buddy's 25 feet back there. He can't see, and we'd lost one way of two-way com. There's not a thing you can do until you get daylight. So it came up daylight. He could see it was daylight. I said, 'Okay, Gene. If this doesn't burn off fast, we're going to call it quits and get out of there.' So after five or ten minutes, nothing happened. So I said, 'Okay. Call it quits, Gene. Get out of there.' He couldn't see. He was absolutely blind and 145 feet away. [...] I started reeling in the tether. [...] He grabbed a hold of it. I said, 'Just walk hand over hand.' So he walked hand over hand, blindfolded. Then I kept pulling. I had this big 125-foot snake in the Gemini cockpit. I kept trying to get it down. [...] I said, 'Look. Take one of your hands and pull down on the helmet as much as you can and put your head up and see if you can take your nose and rub a hole on it.' So he did that, and he could find a little hole he could see. [...] My main thing is to get him in before the next sunset. [...] So he came in closer, and I just reached over and grabbed this over-the-center [door] mechanism and slammed it. [...] So finally he got back in his seat, raised his visor, and his face was pink, like he'd been in a sauna. He says, 'Help me get off my gloves, too.' So I helped him get his gloves off, and his hands were absolutely pink. So I took the water gun and just hosed him down. You shouldn't squirt water around in a spacecraft. Turns out he lost about ten and a half pounds in two hours and ten minutes outside. [...] We landed the next day. [...] Finally, after we did splash down, after the final thing, we're back in the crew quarters having a drink, I told him what Deke had said.

The next mission, Gemini 10, with its crew of John Young and Michael Collins, also had its problems. They also could not see during the spacewalk towards the Agena. Collins says:

The method that we used on Gemini 10 to purge the system, to absorb the exhaled carbon dioxide from your body, were canisters of lithium hydroxide. The stale air went through the lithium hydroxide and it came out purified. Lithium hydroxide is kind of, I think, a granular sort of material, and our best guess was that somehow lithium hydroxide had escaped from some canister and had gotten into the nooks and crannies of the system in the pipes and that there was some triggering mechanism having to do with depressurizing the spacecraft that caused that lithium hydroxide to start billowing up. It went through, and it can be an irritant, and that's what it was. But to the best of my knowledge, they never established that beyond the shadow of a doubt. All I know is that I couldn't see and John couldn't either, and it was frightening for a moment, because the hatch on Gemini was not a very straightforward thing. In other words, you just didn't go 'clunk, latch'. I mean it was – you had to look up, and there were little levers and whichnots that had to be fiddled with, and then you had to make sure that all your hoses and stuff were not going to get in the way, and then you had to come down in a certain way and you had to get your body underneath, your knees underneath the instrument panel and kind of ratchet your body down, and it was a tight fit. So it was the kind of stuff that, with practice, you found it became easy to do, but it was visually dependent [...] it wasn't something you ever had trained to do or thought you would have to be doing anything by feel alone.

By the time we got to Gemini 12 with Buzz Aldrin there were handholds and work stations. On the Shuttle, you see it in space. I mean, they don't go out without being anchored in two or three different ways. But we were stupid; we hadn't thought of that. So, the point is, I was going over to this Agena, propelling myself with this dorky little gas gun. So anyhow, I was using this little gun to get over to something, to grab something that had not been designed to be grabbed, and I'm in this bulky suit that [...] doesn't want to bend too well, I'm immobilized. I'm having a tough time as I'm going along, pitch, rolling, and yawing, trying to keep this dorky little gun through the centre of mass of my body, and then I arrive at this goddamned Agena, which is not meant to be grabbed, and I've got to grab it. So, the first time I grabbed it, I went to the end of it, and it had a docking collar. Docking collars are built to be nice and smooth so that the probe that goes into them will be forced into them. They have smooth lips and edges on them, and that's what I was grabbing. Well, I grabbed the docking collar. Bulky glove, and my momentum is still carrying me along, so I just slipped, and as I went by, then I went cartwheeling ass over teakettle, up and around and about, until I came to the end of my tether, I went back to the cockpit, and then John Young got a little bit closer to the Agena the second time, and when I went over to it the second time I was able to get my hand down inside a recess between the main body of the Agena and the docking collar where there were some wires, and grab some wires.

There were a couple of rendezvous on Gemini 10. We rendezvoused with two different Agenas: our own Agena, call it Agena 10, and then a dead, inert Agena that had been used by the Gemini 8 flight, that had been up in the sky for a couple of months just sitting there. These Agenas were different in two respects. Agena 10 we could ask questions and it would answer. It had a transponder. So we could ask it, 'How far away are you?' and it would tell us. The Gemini 8 Agena had dead batteries. Its transponder could not reply. So when we asked it questions, it would not answer. This meant that we could not find out how far away from the Agena we were. We had to just deduce our range by the apparent size of the Agena or the actual size.

Meanwhile fate had played a hand in getting Buzz Aldrin into space. It was said he was in despair after not being assigned to a prime Gemini crew. He was back up to Gemini 10, which would have meant he would fly on Gemini 13, except that there was to be no Gemini 13. Aldrin was suddenly catapulted to the last mission, Gemini 12, because his next-door neighbour Charles Bassett was killed in a plane crash. Jim Lovell says:

When Gemini 12 rolled around, they said, 'Let's devote a lot of time on 12 to find out how we can really work outside the spacecraft.' And it just so happened that someone, I don't know who it was, said, 'Well, how about under water? Wouldn't that give you sort of an idea of zero gravity if we can make the astronaut neutrally buoyant? And a spacesuit will work just as well under water as it will in space.' So, NASA rented a swimming pool in a boys' school up in Baltimore. Buzz and I went up there, with […] all the crew; and we put Buzz in a spacesuit, got him in the water, made him neutrally buoyant. I had communication with him, and I sat on the side of the pool as if I was inside a spacecraft and we went through some of the basics – we had a crude mockup in the water itself – learning about working in space. Learning about the proper handholds and the proper toeholds to make sure that everything would work. And, of course, on [Gemini] XII, Buzz completed three spacewalks of about five and a half hours, I guess; and, you know, everything was fine.

From his superb walk in space on his Gemini mission, Aldrin went on to be backup for Apollo 8, which paved the way for Apollo 11.

The Gemini project closed in November 1966. It had accumulated 80 man-days in space over ten missions. It had performed orbit changes, spacewalks, rendezvous and re-boosts. It had seen adversity overcome in orbit. The way to the Moon was open. But 1967 was to be a very bad year for everyone.

Despite what they said later the Russians were desperate to beat the Americans to the Moon. They had three manned programs on the go. The first was a manned flight around the Moon called the L1 project. The second – the L3 project – was a manned landing that required a lunar lander and a giant booster called the N1. The third was their orbital missions. For all three the Soyuz spacecraft was the centrepiece.

Its first mission was a planned spectacular, the docking of two Soyuz spacecraft in Earth orbit, followed by the transfer of two crew members from one vehicle to the other via a space-walk. Soviet space leaders believed that this one mission would overshadow the achievements of Gemini. Since September 1965, four Air Force cosmonauts had been training for the com-mander's spot on the two Soyuz spacecraft: veterans Bykovsky, Gagarin, Komarov, and Nikolayev. Vladimir Komarov was the leading contender for the commander of the 'active' Soyuz that would be carrying out most of the manoeuvres.

The Russians could only watch as on 25 May 1966 Apollo-Saturn 500-F – a test vehicle built by Marshall Space Flight Center that duplicated everything except engines and space-craft, of which it had none – rolled out of the VAB on the crawler and moved at glacial speed to Pad A, gleaming in the brilliant sunshine. It was the biggest rocket ever built by man, dummy though it was, and up there, safely on the pad, it was something to behold. It was five years to the day since President Kennedy had proposed landing a man on the Moon and return-ing him safely to Earth.

Engineers began the ground testing of the first flight model of the Soyuz spacecraft on 12 May 1966. There were many problems: over 200 known faults. Instead of the antici-pated 30 days, it took four months to debug the ship and even then the cosmonauts had no confidence in it. There were severe problems with the Soyuz's parachute system. Two of the seven drop tests from an aircraft failed. Kamanin wrote in his diaries:

> One has to admit that the Soyuz parachute system is worse than the parachute system of the Vostoks and the spacecraft isn't much to look at in general: the hatch is small. The communications equipment is out-dated, the emergency rescue system is primitive and so on. If the automatic docking device turns out to be unreliable (which cannot be ruled out) our space program will be headed for an ignominious failure.

The first Soyuz test spacecraft lifted off successfully from Baikonur on 28 November 1966, but entered a lower orbit than expected. The Soviet news agency TASS designated the spacecraft Kosmos-133 and did not indicate that the flight had any connection with the manned space program. The mission ran into problems straight away, making the spacecraft unusable. The second test Soyuz with which it was to have automatically docked was scrubbed.

Two weeks later they tried again. As the rocket ignited they noticed that it didn't seem to be working properly. It shut down and remained on the pad. Steam filled the area as thousands of gallons of water poured onto the launch mount. About 27 minutes after the abort, the launch escape system suddenly ignited. Within seconds the rocket's third stage had caught fire. Kamanin described the scene:

I ran to the cosmonauts' house and ordered everyone who was there to quickly go from the rooms into the corridors. It proved to be a timely measure: within seconds a series of deafening explosions rocked the walls of the building which was located 700 meters from the pad. Stucco fell down and all the windows were smashed. The rooms were littered with broken glass and pieces of stucco. Fragments of glass hit the walls like bullets. Clearly, if we had remained in the rooms a few seconds longer we would all have been mowed down by broken glass. Looking out through the window openings I saw huge pillars of black smoke and the frame of the rocket devoured by fire.

Komarov had been selected for the flight but he knew all too well the problems with the Soyuz. He confided to a colleague, 'I'm not going to make it back from this flight. If I don't make this flight, they will send the back-up pilot instead. That's Yuri, and he'll die instead of me. We've got to take care of him.' There are some reports that Gagarin tried to get Komarov removed from the flight knowing that it would then have to be cancelled because they would not risk him on such a mission.

Gus Grissom, Ed White and rookie Roger Chaffee had been chosen for the first manned Apollo flight and on 27 January 1967, at Pad 34 at Cape Kennedy, all three were performing a so-called 'plugs-out' test of the newly designed Apollo capsule, which was positioned above a Saturn 1B rocket. They were on their horizontal couches, sealed inside the capsule, breathing a high-pressure 100 per cent oxygen atmosphere.

The Command Module, dubbed Spacecraft 12 by its manufacturers, had a history of problems and had arrived late at the Cape. It was four weeks to the launch. The test was long and full of problems. Engineer John Moore said they were performing lots of tests and not really looking at the results. They were working very fast. As the astronauts were being suited up for the test, their secretary, Lola Morrow, sensed a tension and a weariness. They didn't want to do it. Their attitude was 180 degrees off what it was usually, she said.

Visors down, they reached T-10 minutes in the simulated countdown. They were running through a checklist when a voltage spike was recorded at 18.30.54. Ten seconds later Chaffee said 'Hey' and scuffling sounds were heard. Grissom shouted 'Fire', followed by Chaffee, 'We've got a fire in the cockpit', and then White repeated, 'Fire in the cockpit'. Seconds later Chaffee yelled, 'We've got a bad fire! Let's get out! We're burning up! We're on fire! Get us out of here!'

Technician Gary Propst could see Ed White on his monitor. He was trying to open the CM's heavy, two-piece hatch. White had to use a ratchet to release six bolts. He barely had chance to begin loosening the first bolt. In normal circumstances it would have taken about 90 seconds. But with the heat, the fire and the noxious gases accumulating rapidly it was just impossible.

Only 21 seconds after the first indications of fire the transmission ended with a scream. Within seconds the pressure inside exceeded Apollo 1's tolerance and the capsule ruptured. The surrounding area filled with thick smoke. Pad leader Don Babbitt leapt from his desk, shouting at lead technician Jim

Greaves to get them out, not realising they were already dead. 'The smoke was extremely heavy,' Babbitt said, 'heavy thick grey smoke, very billowing, but very thick.' None of the pad crew could see more than an arm's length in front of them. Later, 27 technicians were treated for the effects of inhalation. Fire-fighters eventually opened the hatch.

Then, almost as quickly as it started, the fire was out.

Deke Slayton and flight surgeon Fred Kelly arrived at Pad 34 within minutes. There was an additional problem. The heat might trigger the Saturn rocket's escape tower. With the capsule still hot, the pad was cleared. It was six hours before the bodies were removed. Their flight suits were almost intact, not even blackened. They had all died from asphyxia when their oxygen hoses burned and their suits filled with poisonous smoke. Deke Slayton later described it as the 'worst day' of his career.

Investigators discovered that the fire began under Gus Grissom's seat, on the left side of the cabin, somewhere in the 30 miles of wiring. To this day no one knows what caused the initial spark. The inquiry found that the documentation was so poor that no one was even sure what was within the spacecraft at the time of the accident.

Frank Borman will never forget that night:

We were having dinner with some friends on a lake in Huntsville, Texas; and a highway patrolman knocked on the door and said that I was supposed to call Houston right away. Susan and I left and drove back to Houston and went over to Ed White's house, because Susan was close to Pat White.

Neil Armstrong was in Washington where the President was signing the Outer Space Treaty with other nations that kept the Moon as the property of all people. He said later:

> I'd known Gus Grissom for a long time. Ed White and I bought some property together and split it. I built my house on one-half of it, and he built his house on the other. We were good friends, neighbors. I suppose you're much more likely to accept loss of a friend in flight, but it really hurt to lose them in a ground test. That was an indictment of ourselves. I mean, it happened because we didn't do the right thing somehow.

Walter Cunningham, of the third astronaut intake, called 'the 14', was in training for a forthcoming Apollo flight:

> We knew that there were a lot of problems on the test. We'd run the test the night before without closing the hatch. With any new vehicle and particularly with a new spacecraft there are lots of birthing pains on getting the systems to work. Not just the spacecraft, but getting the means to check out the spacecraft. So, we were waiting. It was a Friday night, and we were all going to come back home for the weekend because we'd been spending a whole lot of time – they [NASA] actually thought that it was within a month of launch. We never actually believed that at all. There were just too many things wrong. But we were going through the fiction of having a scheduled late February launching for Apollo 1.

When Gene Kranz came off Gemini and turned his attention towards Apollo he was in for a big surprise:

> I was really shocked by how far we had yet to go before we could pull together a coherent Apollo operation with the same quality that we were now experiencing in the Gemini operation. And this was particularly true in our relationships with North American Rockwell who made the Apollo Command Module. Rockwell is a very good contractor, but they hadn't been flying in space before. All of our experience had been with McDonnell. And Rockwell was used to building fighter airplanes, rolling them out of the factories, etc., and they weren't about to listen to anybody that wasn't a test pilot. This friction in January, I think, led to the disaster that we had with the pad fire. The fact is that we really weren't ready to do the job, and yet we were moving on. We were sitting there that day, running the test. I had done the shift prior to the fire, and things weren't right that day, and I knew they weren't right. And yet I continued on. I think everybody that was working that test knew things weren't right. We weren't ready! But nobody stood up and assumed the account-ability and said, 'We're not ready. It's time to regroup.' And I think this was one of the very tough lessons that came out of Apollo 1, that we said, 'From now on, we are forever accountable for what we do or what we fail to do.'
>
> We did have an on board tape that was probably run-ning. It was probably burnt up. But, where they had the air-to-ground, which was from the spacecraft to the

blockhouse, and in the 21 seconds from the time you first heard the noise to the time it was over, no one was exactly sure what they said. So, I remember the very next day (I think it was the next day) I was down at the Cape. We flew back either […] Saturday or Sunday. I flew back once, I think, to take Gus's uniform down there for the burial. And then I flew back and we stayed there – Donn and I stayed there – and we sat down with the tapes. We had to – I think Frank Borman and Donn Eisele and I, because we knew the guys, we knew their voices, we sat down and went through this. And even then, we couldn't agree exactly on what went on. And they wanted to get it down to timing. So, I ended up taking those tapes up to Bell Labs up in New Jersey, where they broke it down, did all the magic things they do with it. You know, today it would've been easy with digitizing, but it was tough in those days and they still had some controversy afterwards.

Cunningham says:

It makes the hair stand up on the back of my neck as I think about it now because, you know, […] it's screaming, 'Get us out of here! […] We've got a fire! […] We're burning up! We're burning up!' And aviators […] mostly anticipate that they're going to go either in a big crash, where there's nothing, or that they crash in a fire and they burn. So, fire is really one of the kind of horrors in an aviator's life. He doesn't mind going fast, but he really doesn't want to suffer.

McDivitt says:

> We were doing those same things in Gemini and Mercury.
> We could've had exactly the same problem with Gemini
> and Mercury. We were pressurizing the spacecraft at five
> psi over atmospheric, which was 20 psi. We had a 100 per
> cent oxygen environment. I did the same test on top of
> the Gemini that they were doing at the time that the fire
> occurred. And we did it on every Gemini spacecraft. I
> think we did it on every Mercury spacecraft, too. To this
> day, nobody knows how the fire started. But we just had
> a lot of bad circumstances come together. And some of
> the North American people maintain to this day that they
> were never told that the spacecraft would ever be tested
> in this configuration. If they didn't know it, they were the
> only people in the whole world that didn't know it. But,
> you know, everybody had their own idea how this was
> going to work, I guess. But it was one of those circum-
> stances. You had all this flammable material in there and
> a 100 per cent oxygen environment at 20 psi. That's seven
> times more oxygen than we have in this room right now.

Schirra says:

> I was annoyed at the way what became Apollo 1 came
> out of the plant at North American Aviation's plant in
> Downey, California. It was not finished. It was what they
> called a lot of uncompleted work or incomplete tests and
> work done on it. So it was shipped to the Cape with a

bunch of spare parts and things to finish it out. And that, of course, caused this whole atmosphere of developing where I would almost call it a first case of bad 'go' fever. 'Go' fever meaning that we've got to keep going, got to keep going, got to keep going! When my crew did the test that was followed by Gus and his guys, we were in sea-level atmosphere; no pure oxygen. We were in shirtsleeves. And there were things going on I didn't like at all. I was no longer annoyed; I was really pretty goddamn mad! There were glitches, electronic things that just didn't come out right. That evening I debriefed with Apollo Spacecraft Program Office Manager Joe Shea and Gus. And I said, 'If there are any things that go wrong, like a glitch in the electronic circuits and bad sounds, scrub!' Because Gus and his guys were going to do it in pure oxygen and in an environment that's not very forgiving. We didn't realize how unforgiving it was at that point. We'd gone through the same environment with Mercury and Gemini and made it through. Not that I think of it in that way, but that's how I look at it in retrospect. Gus, I can recall saying, 'If I can't talk to the blockhouse, how the hell are we going to go to the Moon with this damn thing?' That's how bad the communications were. He should have scrubbed. He didn't. He was himself involved in 'go' fever.

Cunningham says:

At the time, everybody was being abused by the schedule. President Kennedy had said, 'We'll land a man on the

Moon and return him safely to Earth in this decade. And here it was, it was […] 1967. Time was getting short and schedule was considered God. So, anything that would slow things down, it was really tough to get through. They didn't ignore it, but it just didn't have the same weight as it did before. The managers had 'go' fever. I think that the Apollo 1 fire was really the key to the successes now we had downstream, because it created not only a different working environment; it developed a firmness of mind that I think was essential to making the right decisions.

We fixed a lot of operational things that had been just rejected out of hand. But now they had the time and the money, and all of a sudden the public and Congress was concerned, real concerned about astronaut safety again. So, we fixed a lot of things and were able to fly a much better spacecraft. The one that we actually flew – you've heard me describe Apollo 1 spacecraft as 'a piece of junk'; and it really was as spacecraft go. The one that we flew was almost perfect! I mean, it was just – you couldn't have asked for a better piece of hardware for the first time. So, that happened, and it enabled us to build one success on another and to make it with six months to spare. And most of us believe that if there had not been that Apollo 1 fire, we would've lost some people in orbit and maybe – who knows what would've happened?

NASA Administrator James Webb was under a great deal of stress. Every day the press contacted him to respond to more and more rumours about Apollo 1. Was he hiding something?

George Mueller tried to take some of the pressure off him, Gilruth and von Braun. But the strain was showing on all. Gilruth had difficulty in moving on. Many believe Webb was never quite the same after the accident. He was aware that a new Deputy Administrator, Thomas Paine, appointed by Johnson in January 1968 to sort out Apollo, was probably going to succeed him. He argued with Senator Walter Mondale who had obtained information that there was in existence a 1965 document by the Apollo program director Sam Phillips detailing Apollo's many problems. Testifying before a Congressional committee, Webb said he had never seen it. The controversy spread to both houses of Congress. Finally they concluded that the report should have been shown to Congress but that it hadn't played a role in the Apollo accident. Mondale wrote a minority opinion report accusing NASA, and hence Webb, of being evasive and displaying a lack of candour and a refusal to respond to legitimate Congressional enquiries. Webb was wounded, politically and personally. He was also losing confidence in NASA's senior management team.

After the accident Frank Borman was asked if he thought that NASA would be unable to recover from the disaster. He replied, 'Never, not for one instant.' Neil Armstrong said, 'We were given the gift of time. We didn't want it.'

Apollo is Faltering

On 3 April 1967, NASA 2, a Grumman Gulfstream aircraft, was taxiing for take-off from Washington National Airport. Bob Gilruth and his Deputy George Low were about to return to Houston after meetings in Washington. Just before starting down the runway, the pilot received a message to return to the terminal where the passengers were to wait in the pilot's lounge. Soon Jim Webb, Bob Seamans, George Mueller, and Sam Phillips arrived. Webb came right to the point: Apollo was faltering; the fire that claimed the lives of three astronauts had been a major setback and its consequences were not yet known. Time, he said, was running out on the nation's commitment to land on the Moon before the end of the decade. Webb asked Low to take on the task of rebuilding the Apollo spacecraft, and meeting Kennedy's deadline.

There was a lot to do, as Low later explained:

By April 1967, when I was given the Apollo spacecraft job, an investigation board had completed most of its work. The board was not able to pinpoint the exact cause of the fire, but this only made matters worse because it meant that there were probably flaws in several areas of the spacecraft. These included the cabin environment on the launch pad, the amount of combustible material in the spacecraft, and perhaps most important, the control (or lack of control) of changes.

Apollo would fly in space with a pure oxygen atmosphere at 5 psi, about one-third the pressure of the air we breathe. But on the launching pad, Apollo used pure oxygen at 16 psi, slightly above the pressure of the outside air. In oxygen at 5 psi things will burn as they do in air at normal pressures. But at 16 psi oxygen most non-metallic materials will burn explosively; even stainless steel can be set on fire. In addition, most non-metallic things will burn – even in air or 5 psi oxygen – unless they are specially formulated or treated. Somehow, over the years of development and test, too many non-metals had been included into the Apollo capsule and nobody had realized the growing threat. The cabin was full of Velcro to help astronauts store and attach their checklists. There was a lot of paper as well as a special kind of plastic netting to provide more storage space. The spacesuits themselves were made of rubber, fabric and plastic.

What was especially worrying to the investigators was that in the haste of the project, the control and documentation of changes had not been rigorous. In fact, the investigation board was unable to determine the precise detailed configuration of the spacecraft, how it was made, and what was in it at the time of the accident. Major changes had to be made, and fast. Time was running out. There were only 33 months remaining from April 1967 and the end of the decade. At any chosen landing site the Sun would only rise 33 more times before 1970.

The Command Module hatch was redesigned to open out instead of in. The Command Module was rewired with better insulation. Much non-metallic material was removed. A new insulating coating that would not burn was developed, only to find that it would absorb moisture and become a conductor, so engineers had to invent another one. Pressure suits had to lose their nylon outer layer, to be replaced with a glass cloth; but the glass would wear away quickly, and shed fine particles which contaminated the spacecraft and caused the astronauts to itch. The solution was a coating for the glass cloth.

Max Faget came up with an idea for the pure oxygen problem on the launch pad: launch with an atmosphere that was 60 per cent oxygen and 40 per cent nitrogen, and then slowly convert to pure oxygen after orbit had been reached and the pressure was 5 psi. The 60:40 mixture was a balance between medical requirements (too much nitrogen would have caused the bends as the pressure decreased) and flammability problems on the other. It worked.

Away from the Command Module, engineers set off small explosive charges inside the burning rocket engines, and to their horror found the LM ascent engine was prone to catastrophic instability – a way of combustion that could destroy the engine on take-off and leave the astronauts stranded on the Moon. A new engine was built, faster than anyone thought possible. Mock-ups of the entire spacecraft were built and set on fire. If they burned, they were redesigned, rebuilt, and tested again. Items were overstressed and overloaded until they broke, and if they broke too soon, they were redesigned, rebuilt and tested again.

After the fire Gene Kranz called a meeting of his staff and said;

> From this day forward, Flight Control will be known by two words: Tough and Competent. Tough means we are forever accountable for what we do or what we fail to do. We will never again compromise our responsibilities. Competent means that we will never take anything for granted. Mission Control will be perfect. When you leave this meeting today you will go to your office and the first thing you will do there is to write Tough and Competent on your blackboards. It will never be erased. Each day when you enter the room, these words will remind you of the price paid by Grissom, White and Chaffee. These words are the price of admission to the ranks of Mission Control.

While NASA was coming to terms with the Apollo 1 disaster, the third unmanned test flight of the Soyuz spacecraft took place. TASS announced the flight as Kosmos-140, another in a long series of nondescript satellites with no stated mission. On 7 February it took off and this time reached orbit successfully. Trouble started on its fourth orbit. The solar panels were not being pointed towards the Sun so that its batteries could be charged. In addition, fuel levels in the manoeuvring thrusters were at 50 per cent far too soon. There were further malfunctions including a too-steep re-entry that breached the heat shield. Any crew on board would have been killed. Remarkably, it was believed that all of its problems could have been overcome if a cosmonaut were on board. The remaining systems, such as life support, the main engine, and thermal control, worked well. Nevertheless, the cosmonauts could see that the Soyuz capsule was a reckless gamble. It was not ready, but officials kept pointing out that a cosmonaut had not been in space for nearly two years.

The State Commission decided to press ahead with the dual manned launches, setting 23 April as the launch date. In one Soyuz would be Komarov. The following day, as the Soyuz was flying over Baikonur, Soyuz 2 would be launched with Bykovsky, Yeliseyev and Khrunov on board. After docking, Yeliseyev and Khrunov would spacewalk to the Soyuz 1, which, with a crew of three, would return the following day. Soyuz 2, now with a crew of one, would also return that same day.

There exists a film of Komarov being driven to the launch pad in a bus. He looks condemned. Kamanin and Gagarin accompanied him to the rocket: Gagarin went up with him all the way to the top of the rocket and remained there until the hatch was sealed. Soyuz-1 lifted off on time. Komarov was the first cosmonaut to make a second flight. He was 40 years old.

He ran into problems straight away. One of the two solar panels did not deploy, resulting in a shortage of power for the spacecraft's systems. Komarov also had problems orienting the Soyuz. By orbit 13 the stabilization systems had failed completely. Unconfirmed reports suggest that Komarov even tried to knock the side of the ship to jar open the stuck panel. Due to dwindling power, he could not stay in space for long. The second Soyuz flight was cancelled and plans made for an emergency re-entry. The three cosmonauts of Soyuz 2 pleaded to be allowed to launch: perhaps they could perform a spacewalk to free the jammed solar panel on Soyuz 1? But they were turned down.

Chertok carefully checked the set of instructions that Gagarin personally transmitted to Komarov. There would shortly be a break in communications as the capsule entered an ionization layer. In the final seconds before loss of contact, Mishin and Kamanin both wished Komarov good luck.

At the appointed time, the re-entry rocket did not fire. Communication with Komarov was re-established and the problem rectified. Another attempt was made. Komarov did not have many more chances left. Miraculously Komarov manually oriented the Soyuz and performed the de-orbit burn. About fifteen minutes after retrofire there came another break in communications. A few minutes later, Komarov's voice cut through the radio silence, sounding 'calm, unhurried, without any nervousness'.

The pilot of one of the search and rescue helicopters fly-ing east of Orsk reported that he could see the Soyuz capsule. When they reached the landing site, it was clear that there had been a disaster. The re-entry capsule was lying on its side, and the parachute could be seen alongside. It was surrounded by clouds of black smoke. The capsule was crushed, on fire and completely destroyed. The parachutes had not worked. Komarov's body was found in the crushed capsule. Medics gath-ered what they could and a few days later his ashes, like those of Korolev, were interred in the Kremlin Wall after a state funeral.

But so great was the destruction of the capsule that later a group of Young Pioneers found more remains that were later buried at the crash site.

To add to the problems with the recovery from the Soyuz disaster, Premier Brezhnev wanted a space spectacular to coincide with the 50th anniversary of the October Revolution in November 1967 – preferably a cosmonaut trip around the Moon. Indeed, just days after Komarov's death, Chief Designer Mishin set out a new plan for the circumlunar project that envisaged four automated spacecraft flying around the Moon between June and August 1967, followed by three piloted flights to make the November 1967 political deadline.

Many in the West believed that the Moon was the Russians' goal. In May 1967 Gemini astronauts Michael Collins and David Scott visited the Paris Air Show at the same time as Soviet cosmonauts Pavel Belyayev and Konstantin Feoktistov. It was only a month after Komarov's death and the Americans gave their condolences. What they were told was that there would be several Earth-orbital test flights that year followed by a circumlunar flight. Collins later recalled that Belyayev said he expected to make a circumlunar flight in the not-too-distant future. Later that year Academician Obraztsov said that 'the very next milestone in the conquest of space will be the manned circumnavigation of the Moon, and then a lunar landing'.

In August there was a brief news report that ten Soviet cosmonauts were practising splashdown tests for future space missions. It was significant because unlike Earth-orbital missions, cosmonauts returning from the Moon would land on water.

It was clear that due to the accumulating problems the Soyuz docking and spacewalk mission, conceived as practice for later missions in which the crew would transfer to a lunar transit craft – the L1 – would be delayed, so the plan was changed. There would be no transfer. Cosmonauts would launch in the L1 direct on a more powerful rocket – the Proton. Because of this, two additional automated circumlunar missions were put into the flight sequence, making a total of six robotic flights before a piloted one. If they all went well then there was a chance they could fly a man around the Moon by the November 1967 political deadline.

Once again it was a highly dangerous strategy. Testing on the spacecraft had hardly started but even in the early stages had been beset with poor standards by contractors. Communist Party and government leaders were anxious knowing that the first launch of the Americans' mighty Saturn 5 rocket was due for late 1967 while their own giant N1 booster was still many months away from launch. In August 1967, Secretary of the Central Committee for Defence and Space Ustinov was infuriated. He told Mishin: 'We have a celebration in two months, and the Americans are going to launch again, but what about us? What have we done?' As the summer wore on some degree of sense prevailed. It was realized that the November deadline was impossible. With the pressure off for a little while, Mishin could think about beating the Americans in a flight around the Moon. Following the Apollo fire, US manned flights would not take place until mid-1968 at the earliest. It gave them breathing space.

Meanwhile the manned Soyuz flights in Earth orbit were continuing. Two automatic Soyuz capsules were prepared to practise rendezvous and docking manoeuvres. The first launched and, for the first time in the Soyuz program, all its systems were functioning without fault when it reached orbit; the solar panels deployed, and the Igla ('Needle') radio docking system seemed to be working. On the second day of the flight there were some glitches but the State Commission gave the go-ahead for the second Soyuz launch. It was launched on 30 October. Within 62 minutes of its launch both vehicles were docked – the first automated docking ever performed. After they were linked, however, controllers discovered that there had not been a 'hard' docking because there was a gap between the two ships. Upon analysis this was considered a minor problem, and after three and a half hours they separated.

At the same time, the Americans achieved a major milestone on their more logical and organized attempt on the Moon. Apollo 4 was launched on 9 November – the first test, unmanned, of the Saturn 5 rocket. Eighty-nine trucks of liquid oxygen, 28 of liquid hydrogen and 27 rail cars of RP-1 (refined kerosene) were delivered to the Cape. The Saturn's first stage, consisting of five mighty F-1 rocket engines, burned RP-1 and liquid hydrogen, providing 35 tons of thrust; the second stage comprised four powerful J-2 engines using liquid oxygen and hydrogen; and the third stage used a single J-2. All worked flawlessly. Von Braun watched the rocket rise into the sky. Apollo 4 subjected the CSM to the lunar return speed of over 40,000 km per hour. Temperatures on the heat shield reached 2,750°C, more than half the surface temperature of the Sun. The heat shield charred as expected, but the inside of the cabin remained at a comfortable 21°C.

In January 1968 the L1 cosmonauts began training in a specially built simulator, carrying out at least 70 runs. The following month Mishin and Kamanin agreed on the selection of four commanders for the first few missions: Bykovsky, Leonov, Popovich, and Voloshin. They did not have much confidence in the spacecraft. Kamanin recalled that by March the cosmonauts were working diligently and knew the craft well: 'Perhaps, it is precisely because the cosmonauts excellently know all the strong and weak points of the craft and the carrier rocket that they no longer have their initial faith in the space hardware.'

In early 1968, Mishin and Chelomei agreed to launch a test spacecraft out to a distance of about 330,000 km into deep space – lunar distance – and bring it back to Earth, simulating an actual circumlunar flight. But there was an air of pessimism, a feeling of inevitable failure that was difficult to shake off. Kamanin wrote in his diary: 'All of us need a successful launch like a breath of fresh air. Another failure would bring innumerable troubles and may kill the people's confidence in themselves and the reliability of our space equipment.'

A Minor Mutiny

Apollo 5 on 22 January 1968 was the first flight test of the Lunar Module, unmanned, in Earth orbit. There were problems. The computer shut down the LM's descent engine prematurely on its first burn. Flight controllers took over and continued with an alternative mission. Now another question arose: should they repeat the flight? After considerable technical debate, it was decided that the next flight with the LM would be manned – which it was, fourteen months later. Apollo 6, three months after Apollo 5, was to be a simple repeat of Apollo 4, but it wasn't.

For the first two minutes of Apollo 6's flight, the five F-1 engines on the first stage burned normally, but then experienced peculiar thrust variations, like a pogo stick, lasting about 30 seconds. Then when the second stage was operational and four minutes into a six-minute 'burn', two of the five engines shut down, requiring the others to fire for an additional 59 seconds to compensate. After a difficult ascent, during which the upper stages were at times flying backwards, it ended up in the wrong orbit. As if that wasn't enough, the upper stage fired to reignite for its second burn to place it into a higher orbit. 'If this had been a manned flight,' wrote Slayton in his memoir, *Deke*, 'the escape tower on the Apollo would have been commanded to fire, pulling the spacecraft away from the Saturn for a parachute landing in the Atlantic.'

By March 1968, NASA had still to fully recover from the Apollo 1 tragedy and was months away from flying a piloted Apollo spacecraft in Earth orbit, let alone in lunar orbit. Many Soviet officials believed that it would take a miracle for the Americans to successfully carry out a sequential series of successful Apollo missions in the months leading to a first landing by the decade's end. But in many ways, the Russians were viewing American capabilities like their own. Failures were an accepted part of the Soviet space effort, more so than in the US.

In a diary entry in March 1968, Kamanin wrote:

It took us three extra years to build the N1 and the L3, which let the United States take the lead. The Americans have already carried out the first test flight of a lunar spacecraft, and in 1969 they plan to perform five manned flights under the Apollo program.

Their test Moon ship lifted off on 2 March 1968, into a circular Earth orbit. Soon afterwards the Block D booster stage fired for 459 seconds to put it into a highly elliptical orbit with a high point of 354,000 km – lunar distance. TASS did not announce anything of note about the launch, except to name the spaceship Zond 4 (*zond* being the generic Russian word for 'probe'). At the end of its mission it evidently passed through the atmosphere safely and was about to deploy its parachutes near the West African coast when the emergency destruct system on the descent capsule ignited. A destructive charge had been included on the spacecraft for fear that the Americans may get hold of it. Another test flight took place the following month. This time the third stage failed to ignite and the emergency rescue system was activated. The political leadership was extremely worried by the accumulating series of failures in the program. Despite them, Mishin was ordered to accelerate the pace of work on the L1 to launch a crew around the Moon by October 1968.

A test flight around the Moon was scheduled for July but days before, as the rocket and spacecraft were being tested on the pad, the Block D stage suddenly exploded, killing one person. The aftermath of the accident was extremely dangerous. Observers watched in terror. The lunar spacecraft and part of Block D were balanced, ready to fall and explode at any time. Engineers risked their lives dismantling the explosive wreckage. The July and August lunar launch windows were abandoned.

On 6 May Neil Armstrong came close to being killed. He was flying the Lunar Landing Research Vehicle (LLRV) at Ellington Air Force Base near Houston. This was basically a single jet engine pointed downwards. Called the 'flying bedstead', it was notoriously tricky to handle. On a simulated lunar descent, leaking propellant caused a total failure of his flight controls and forced Armstrong to eject. Alan Bean, one of the intake of astronauts after Armstrong, saw him that afternoon back at his desk in the astronaut office. Bean then heard colleagues in the hall talking about the accident, and asked them, 'When did this happen?' About an hour ago, they replied. Bean returned to Armstrong and said, 'I just heard the funniest story!' Armstrong said, 'What?' 'I heard that you bailed out of the LLRV an hour ago.' 'Yeah, I did,' replied Armstrong. 'I lost control and had to bail out of the darn thing.' 'I can't think of another person,' Bean recalls, 'let alone another astronaut, who would have just gone back to his office after ejecting a fraction of a second before getting killed.'

In the summer of 1968 the US press was full of rumours about the impending launch of a super-heavyweight Soviet rocket comparable to the Saturn 5. The media in the West did not know the details about how badly it was going for the Russians. James Webb, said: 'there are no signs that the Soviets are cutting back as we are. New test and launch facilities are steadily added, and a number of spaceflight systems more advanced than any heretofore used are nearing completion.' Later, George Mueller wrote in a memorandum distributed to Apollo contractors that the Russians were developing a 'large booster, larger by a factor of two, than our Saturn 5'.

As summer gave way to autumn the piloted circumlunar program was getting into deeper trouble. In four tests of the L1 spacecraft since late 1967 there had been three complete failures and one partial success – the mission of Zond 4 in March 1968. Another L1 spacecraft had been destroyed during ground preparations for a test launch in July 1968. It was under this cloud that the first of the three remaining L1 spacecraft arrived at the Baikonur Cosmodrome for a new series of attempts beginning with the lunar launch window in

September. This time the L1 launch was perfect. The Proton booster lifted off on 15 September with the Moon suspended tantalisingly above the pad. At an altitude of 160 km, the third stage ignited and after a tense 251 seconds the stack went into a perfect Earth orbit of 191 by 219 km. After a circuit around Earth the Block D fired a second time to send it towards the Moon. Shortly afterwards the Soviet press announced the launch, designating the mission Zond 5. It was the first time in their circumlunar program that a spacecraft had been successfully sent towards the Moon. Engineers and cosmonauts were jubilant. A few days later it circled around the Moon at a distance of 1,960 km and was flung onto a return trajectory towards Earth. It splashed down on 21 September in the Indian Ocean, where it was hauled in by the Soviet Navy – watched closely by the US Navy. A pair of tortoises were on board, among other animals, and they survived their ordeal, becoming the first earthlings to go to the Moon.

Those responsible for the Russians' large booster – the N1 – knew it was a gamble. The so-called Council for the Problems of Mastering the Moon met on 9 October to discuss the overall status of the Soviet lunar landing program. Mishin said that the first N1 flight model would only be able to lift a disappointing 76 tons but a modification of the second stage would allow the attainment of the 95 tons needed for a lunar landing by a single cosmonaut. More improvements might make it possible to take two cosmonauts to the Moon's surface. Academy of Sciences President Keldysh was one of the strongest supporters of the two-cosmonaut plan, considering sending just one cosmonaut very risky; but then he made the somewhat reckless proposal that they should seriously consider landing two cosmonauts to the Moon on the very first launch of the N1. If that was impossible, then the mission should be to land a lone cosmonaut. Brezhnev was keen for some success and is reported to have said: 'We should prepare for a manned mission to the Moon straight after the first successful launch of the N1, without waiting for it to be finally developed.' Brezhnev's demands emphasize the gap between the people building the spacecraft and those who controlled the money. One could say that the politicians did not understand the true engineering facts of the situation, but then again, the engineers themselves seemed to be turning a blind eye towards them.

The Zond 5 mission was the first real success in the L1 Moon program. It allowed the USSR to plan on flying a crew on a circumlunar mission in January 1969, dependent upon two more successful L1 flights, even though many realized that was wishful thinking. The cosmonauts had almost completed their training program and it was hoped that one of the crews would make history as the first humans to fly from Earth to the Moon. But the men training for a circumlunar mission were not the only cosmonauts preparing for spaceflight in the fall of 1968. By August, cosmonauts Beregovoi, Volynov, and Shatalov had completed their preparation for the first piloted Soyuz mission since the Soyuz 1 tragedy more than a year before.

More by luck than planning, the Soviet 'return to flight' Soyuz mission would take place in time for the 51st anniversary of the October Revolution. The idea was to carry out a manned repeat of the successful automated docking of a year before; that is, for one cosmonaut in an active Soyuz to link up with a passive automated Soyuz. The two ships would remain docked for a few hours before separating and carrying out independent missions. Such a conservative rendezvous and docking mission would hopefully lead the way for the long-delayed inter-ship cosmonaut transfer attempt. The Soviet political leadership was anxious to resume space missions after the long gap, particularly because of NASA's forthcoming Apollo 7 mission in October – the first manned US spaceflight since the Apollo 1 fire in January 1967.

Consider the mood of America as it approached the end of 1968, by any accounting one of the unhappiest years of the 20th century. It was a year of riots, burning cities, sickening assassinations, and universities forced to close their doors. In Vietnam the twelve-month toll of American dead rose to 15,000, and the cost of the war topped $25 billion. By mid-December the country's despair was reflected in the Associated Press's nationwide poll of editors, who chose as the two top stories of 1968 the slayings of Robert Kennedy and Martin Luther King; *Time* magazine picked a generic symbol, 'the dissenter', as its Man of the Year. The poll and the Man of the Year were scheduled for year-end publication.

Webb left NASA on 7 October. He had several reasons, some of them personal, some of them strategic. He had been in the job for eight years and it had taken its toll; however, the next year could see the Moon landing and it would be a fine climax to his role if he was the administrator when it happened. On the other hand there were political issues. Johnson had told him that he was not running for president in 1968 and protocol dictated that on the arrival of a new president, all heads of agencies like NASA offered their resignations. But the first manned Apollo flight was coming, as well as Apollo 8's voyage around the Moon. Webb felt that if anything went wrong he didn't want to be a defensive administrator but have the freedom outside NASA to defend it. It would also give whoever won the presidential race – Richard Nixon or Hubert Humphrey – a clean sheet should there be any loss of life. As expected, Thomas Paine, took over as acting administrator.

The redesigned Apollo capsule was launched as Apollo 7 on 11 October 1968 from Pad 34 at Cape Kennedy. On board were Wally Schirra, Donn Eisele and Walter Cunningham – the Apollo 1 backup crew. It was not only the United States' return to flight after the tragedy but an important shakedown flight to test the cone-shaped Apollo Command Module for the first time in space, along with its associated Service Module. It was also the first manned flight of a Saturn booster; in this case, the Saturn 1B variant. Sixty-eight metres tall, men had never ridden into space on a more powerful rocket.

Schirra was now 45 years old and making his third spaceflight. Alongside him were two rookies. Knowing that it was almost certainly his last trip into space, Schirra was determined that it should be a perfect mission and especially Schirra's mission. Unfortunately, shortly after lift-off he developed a cold.

The mission was to include a television broadcast from the Command Module. Schirra tells the story:

> We launched on a Friday. I remember this very specifically. In orbit, our so-called Friday night, Donn Eisele was on watch and Cunningham and I were supposed to be

sleeping. And I hear Donn saying, 'Wally won't like that.' I put on my mike and listened in. 'Oh, we're supposed to put on the television tomorrow morning.' I said, 'Well, we didn't have it in the schedule, gentlemen. That doesn't go on till Sunday morning.' I should have said, 'I don't want to interrupt *Howdy Doody* [a popular TV program]', but I wouldn't have gotten away with it. What I really was saying, 'We have not checked this system out. It's in the flight plan to be checked at this point in time. We'll check it at that point in time.'

We did *The Wally, Walt, and Donn Show* Sunday. But by then, everybody was saying, 'These guys are getting testy up there. They're not mutinied, but they're not going along with the flight controllers.' I have yet to meet a flight controller that ever died falling out of a chair. That was my whole attitude from then on. 'Don't mess with me, guys! This is my command', and I wasn't kidding. And, 'I'll take all the advice, all the information you can give me, but don't push us around.' We're still worried about whether this is a safe spacecraft or not. We had even gotten to the point where they were going to shave all our hair off in case there was a fire. And why am I going to start running a TV show for somebody if I haven't checked the camera out, all the electrical circuits, piece by piece? Ah-hah, it works. Now we'll show you TV. Oddly enough, we got an Emmy for that *Wally, Walt, and Donn Show*, so I can't really say it was a bad deal.

Cunningham says:

Well, Apollo 7 became very important. If we had not had a success on Apollo 7, we really don't know what would've happened to the space program. Another accident and the fainthearted in the country, as we have a tendency to be, would've been clamouring to stop it. There was some real bickering back and forth between Wally and the ground. I, frankly, have never felt like I had any kind of a problem with the ground, but Wally was still demonstrating that it was Wally's flight and Wally was in charge. He has maintained since, that he felt the responsibility. He's never said that what he did was anything except the responsible thing to do. I really think it's a case of, on some instances, Wally wanting to insist he was in charge when nobody cared who was in charge anyway.

The mission lasted almost eleven days and was successful. They simulated many of the events that would be required for a mission to the Moon. At one stage their rocket propelled them into a 269-mile-high orbit; 'that was a real kick in the pants,' exclaimed Schirra. Re-entry was normal, although Schirra refused to don his helmet for the procedure. The Apollo equipment received a thumbs-up, even if the commander of the flight didn't. Prior to the flight the Apollo 7 crew did not have a good reputation. According to Collins, Schirra was late to work every morning and never apologized, and never tried to catch up with the schedule. Cunningham 'bitched' all the time and Eisele served as a good-natured referee who didn't quite understand what was going on half the time. After the mission none of them flew in space again. But Wally's flight was an ending rather than a beginning. It marked the end of the slippage in the Apollo program.

Just a few days after Apollo 7 returned, the unmanned Soyuz 2 passed over its launch site and Soyuz 3 lifted off with Colonel Georgi Beregovoi aboard. It was the first ever piloted launch from Site 31, the second launch complex at the Baikonur Cosmodrome and the 25th manned spaceflight. At 47 years old, Beregovoi was the oldest person to go into space at the time. When in orbit the Igla automated docking system brought Soyuz 3 to within 200 metres of Soyuz 2 before Beregovoi took over manual control. But the two ships were not aligned perfectly and instead of stabilising his ship along a direct axis to Soyuz 2, Beregovoi put his spacecraft into an incorrect orientation. This meant that Soyuz 2's radar system, sensing something was wrong, automatically turned its nose away to prevent what it saw as an incorrect docking. Beregovoi did not see the problem and performed a fly-around, and then tried to approach Soyuz 2 for a second time but the same thing happened, by which time he had almost exhausted all the propellant he had available for such manoeuvres, meaning that further docking attempts had to be called off. For three days, 22 hours, and 50 minutes Beregovoi had circled the Earth, completing 64 orbits, and while his flight may not have been successful, at least it was not a disaster.

Because of delays with the next flight-ready L1 vehicle, the Russians had to forgo the October lunar launch window, thus shifting any possible launch into November. Soviet space planners were aware of the rumours of an Apollo lunar-orbital mission by the end of the year so they resorted to their usual public tactic of obfuscation, giving contradictory positions. On 14 October Academician Sedov, who was representing the Soviet Union at the 19th Congress of the International Astronautical Federation in New York, said that the 'question of sending astronauts to the Moon at this time is not an item on our agenda. The exploration of the Moon is possible, but is not a priority.' It was a lie.

The success of the Apollo 7 mission crystallized an audacious idea that had been kicked around at NASA. It was in early August that George Low, the Deputy Director of NASA's Manned Spacecraft Center in Houston, ordered his staff to work on a plan in which an Apollo Command and Service Module launched on a Saturn 5 would go directly to lunar orbit. It was a risky decision as it would be only the third launch of the Saturn 5 booster, the first time men had flown on it, as well as the obvious fact that the risks of a going into lunar orbit were far greater than going into orbit around the Earth. But the advantages were many in technical and scientific knowledge as well as a demonstration of what the United States could achieve. A few weeks later NASA HQ gave its approval for the mission provided that Apollo 7 was successful. Undoubtedly, another reason was the fact that Zond 5 had gone around the Moon and, as far as NASA knew, a Soviet manned circumlunar flight could take place at any time. The Russians had a lunar launch window in December 1968. Would they be able to upstage Apollo 8?

How did Deke Slayton select the men to fly? Mostly by seniority. An astronaut stood in line until his turn came, though the order of assignment within his own group was important. 'They are a durn good bunch of guys, real fine troops, a bunch of chargers,' said Slayton in 1972. 'It's not the kind of organization where you have to keep pointing people in the right direction and kicking them to get them to go. Everybody would like to fly every mission, but that's impossible, of course. They all understand that, even if it makes some of them unhappy.' Slayton was rarely accused of unfairness, a remarkable achievement considering the stakes involved; his own eventual assignment, at age 48, to the 1975 Apollo-Soyuz docking mission (detente in space) was universally popular.

Not all astronauts were equal, of course. Before the first manned Apollo flights, six crews had been formed, commanded by Schirra, Borman, McDivitt, Stafford, Armstrong, and Conrad. Before he flew Apollo 7, Schirra announced he was quitting afterwards, and that left five. Which one would land first on the Moon? It depended on the luck of the draw. 'All the crews were essentially equal,' said Slayton, 'and we had confidence that any one of them could have done that first job.'

At one time either Borman or McDivitt seemed likeliest to be first on the Moon; after they flew their early Gemini missions they were sent straight to Apollo instead of being recycled into later Gemini flights. But in 1968 two things happened to derail this prospect: McDivitt, scheduled to lift off on the first Saturn 5 (Apollo 8), declined the opportunity because it would not carry the LM, on which he had practised so long. Borman, scheduled for Apollo 9, was 'highly enthusiastic' about Apollo 8, LM or no LM, but in deference to wife and children he decided that it would be his last flight.

Borman's backup was Armstrong; McDivitt's was Conrad. In each case, the backup shifted with the prime, so Armstrong in the normal rotation of commanding the third mission after being a backup, became commander of Apollo 11, which was considered to be the first landing mission, and Conrad lost his chance to be first man on the Moon by moving to Apollo 12. There was also the possibility that Apollo 10, commanded by Stafford, might be the first Moon lander because George Mueller initially saw no point in going to the Moon a second time without touching down. But the LM wasn't completely adapted for the task (it weighed too much) and the program management decided that they were not ready for the big step.

Apollo 7's achievement led to a rapid review of Apollo 8's options. The Apollo 7 astronauts went through six days of debriefing for the benefit of Apollo 8, and on 28 October the Manned Space Flight Management Council chaired by Mueller met at Houston to investigate every phase of the forthcoming mission. Next day came a lengthy systems review of the Apollo 8 spacecraft, which had been designated 103. Paine made the go/no go review of the Apollo 8 lunar orbit mission on 11 November at NASA Headquarters in Washington. By this time, nearly all the sceptics had become converts. At the end of the meeting Mueller put a recommendation for lunar orbit into writing, and Paine approved it. He telephoned the decision to the White House, and the message was laid on President Johnson's desk while he was conferring with Richard M. Nixon, elected his successor six days earlier.

'You have bailed out 1968'

The crew of Borman, Stafford and Collins who originally trained together weren't kept together. Tom Stafford was given a new crew of John Young and Gene Cernan and they were assigned to be the backup crew for Apollo 7. Later they would be the prime crew for Apollo 10 – the last mission before the first landing. Frank Borman and Michael Collins were promoted to the prime crew of Apollo 8 along with Bill Anders. But the changes meant that Collins lost his chance to be considered for one of the first Moon landings as he was in his own words 'promoted' from Lunar Module pilot to Command Module pilot. This was because Deke Slayton had a rule that on all flights the Command Module pilot had to have flown in space before. Collins then suffered a slipped disc so Jim Lovell replaced him.

Jim Lovell recalled how he heard he was going to the Moon for the first time:

> We were going to go out to 4,000 miles so that we could test the lunar module, the command module, and then

come back at a high rate of speed so that, you know, we could test the heatshield and things like that. I recall this very vividly. The three of us were out at Downey at North American testing our spacecraft; and Frank got a call to go back to Houston. So Bill Anders and I still stayed out there. We were working out there. And Frank came back again, back to Downey, and said, 'Things have changed.' And we said, 'Viz. a what?' He said, 'If everything goes all right with Apollo 7, we'll – Apollo 8 will go to the Moon.' I was elated! I thought, 'Man, this is great!' I mean, I had already spent two weeks in space in Gemini VII with Frank Borman. I didn't want to spend another eleven days, or something like that, you know, going around the Earth again. I said, 'This is fantastic!'

Gene Kranz says:

Most of the people give the credit for Apollo 8 to a decision in August where George Low came down and said, 'Hey, you know, I think, in order to keep this program on track – we've got problems in the lunar module; it's behind schedule, it's overweight, there are software problems there – I think that we've got to go to the Moon.'

Well, if I go back into the March/April timeframe, Chris Kraft at one of the staff meetings was concerned about the same types of things. And we had what we called an E-mission. It was one where we were just taking and putting a very large 4,000-mile orbit mission into this package. And this was going to test the command and service module and lunar module, but in a very high elliptic orbit. I don't think anybody in the program thought that made much sense, but it was there. So Kraft started playing games with this mission.'

He kept saying, 'Well, Johnny, how big could we make that orbit?' And Johnny Mayer would say, 'Hell, we could make it so big we'd go around the Moon if we wanted to!' And Kraft said, 'Gee, we ought to develop some kind of an alternate for this E-mission.' He said, 'Johnny, why don't you look at it?' So that was in the April timeframe. Come May, Johnny Mayer and – Mayer loved to have work for his conceptual flight planners. You know, the conceptual flight planners were sort of like the mobile strike force; they were sort of the eggheads in a very eggheady division. But boy, the one thing they could do is figure out how to do difficult and complex things in a trajectory sense. In May, they now came back in and provided a series of briefings. And in these briefings now, this 4,000-mile orbit had grown to encompass the Moon! But it still had a Command and Service Module and Lunar Module in it. So now comes June. Kraft says, 'Look, what happens if the LM can't make it? What do we get out of it?' And the obvious one, 'Well, gee we figure out whether all of our navigation works, our tracking works, and all these kinds of things.' The bottom line of this thing was, I really think it was either Kraft planted the seed very strongly in George Low's mind or George Low had some good mission staff engineers that knew what the mission planners were doing. And I think if you really think about it, that's a very short turnaround to come up with such a monumental decision, to have all the data on the table ready to go.

By early November, the Russians were still planning two more automated L1 lunar missions, one in mid-November and one in early December, to be followed by a manned circumlunar flight in January. But once they heard of the Apollo 8 lunar orbit idea they must have realized that they had an advantage if they could but use it. The Apollo 8 launch window opened on 21 December but because of different lunar trajectories undertaken from the two launch sites the circumlunar launch window for a Soviet launch from central Asia would occur earlier, around 8–10 December. But despite much press speculation in the West and an increase in tension approaching 8 December, they were just not in a position to take advantage of the difference. An automated L1 launch did take place in November, sending the spacecraft, designated Zond 6, towards the Moon. As soon as it was on its way, controllers discovered that an antenna boom had not deployed. Despite this, the mission went very well, with Zond 6 flying around the far side of the Moon two days later at a closest distance of 2,420 km. After it had circled the Moon, controllers had to refine its trajectory for it to perform a guided re-entry into Earth's atmosphere and land on Soviet territory. The first correction was successfully accomplished and it looked as if everything was on track until controllers detected a disastrous problem: the air pressure within the descent apparatus had dropped, indicating a compromise of the spacecraft's structural integrity. Despite the partial depressurization, later found to be the result of a faulty rubber gasket, the critical systems on the ship remained operational, and the controllers were able to carry out the third and final mid-course correction, just eight and a half hours prior to re-entry, at a distance of 120,000 km from Earth.

On the morning of 17 November, Zond 6 separated into its two component modules prior to re-entry, passing through its 9,000-km-long re-entry corridor. It skipped out of the atmosphere, having reduced velocity down to 7.6 km per second, and began a second re-entry that further lowered velocity to only 200 m per second. The complex re-entry was a remarkable demonstration of the precision of the L1 re-entry profile designed to reduce G forces. However, during part of the descent, pressure in the descent apparatus reduced further, killing any biological specimens. No doubt, a crew within the ship would have perished as well. Then the parachute system failed and it plummeted to the ground and smashed into pieces. Remarkably, the impact occurred only 16 km from the Proton launch pad at the Baikonur Cosmodrome where Zond 6 had lifted off just six days and nineteen hours previously. Among the items recovered intact from the wreckage was the exposed film from the camera, which provided beautiful pictures of both Earth and the Moon.

Because of the crash, Mishin postponed any plans for a piloted L1 mission in the near future: the dreams of Soviet engineers and scientists of circling the Moon prior to the United States were over. As the historic Apollo 8 launch grew closer, Soviet spokespersons began to neutralize what was undoubtedly a public relations disaster. Veteran cosmonaut Titov, on a trip to Bulgaria, told journalists the day before the Apollo 8 launch,

'It is not important to mankind who will reach the Moon first and when he will reach it – in 1969 or 1970.'

But it did matter. It meant everything.

When Apollo 8 lifted off from Cape Kennedy on 21 December 1968, the eyes of the world were upon the three astronauts who were embarking on a journey as important as any in history – to break the bonds of Earth and head out into deep space. Kamanin wrote in his diary:

> The flight of Apollo 8 to the Moon is an event of world-wide and historic proportions. This is a time for festivities for everyone in the world. But for us, the holiday is darkened with the realization of lost opportunities and with sadness that today the men flying to the Moon are not named Valeri Bykovski, Pavel Popovich, or Aleksei Leonov, but rather Frank Borman, James Lovell, and William Anders.

Borman says:

> I didn't […] want the mission to get fouled up because we really weren't certain that the Russians weren't breathing down our backs. So I wanted to go on time. […] Saturn V was a unique vehicle. And of course, it was powerful and noisy and vibrated, and the stagings were really kind of violent. But when you got on the third stage it was smooth and quiet and was just like the upper stage of the Gemini. Actually, it was less demanding than Gemini from a G standpoint, because it didn't reach the high Gs.

I don't think there was any fear. The main fear was that somehow we'd screw up.

William Anders' first ride into space was the first manned flight of the mighty Saturn V:

I mean it was like violent sideways movement and massive noise that … nowhere near had been simulated properly in our simulations. For about the first ten, seemed like forty, but probably the first ten seconds we could not communicate with each other. Had there been a need to abort detected on my instruments I could not have relayed that to Borman. So we were all out of it for the first ten or twenty seconds.

The next most impressive thing was that as we burned out on the first stage we were hitting about six or eight G's and we were back in our seats. You could hardly lift your arms, you have trouble breathing, but you're not blacked out because the way your blood was flowing from your legs down into your torso. But try to reach up, it's like you had a twenty pound weight in your hand.

Then the engines cut off, and just as they cut off some retro rockets fire to try to move that big first stage away from the second and third stage but slightly before it separates. So, you go from a plus 6 G to a minus one-tenth, and the fluid in your ears just goes wild. I felt like I was being catapulted right through that instrument panel. Instinctively I put my hand up in front of my face and just about the time I got my hand up the second stage cut

in and whack-o, right onto the face plate with the wrist
ring which left a gash. I thought, 'Oh, damn, here I am,
the rookie of the flight, and sure enough here's this big
rookie mark. When we got into orbit and I got out of my
seat and we took off our suits and each guy handed me
their helmet to stow and, sure enough, each one of them
had a gash in it from the same thing.

Entering Earth orbit the Apollo 8 Command and Service
Module and its crew had to fire the main engine that would
take the spacecraft away from the Earth, making them the first
men to ever leave their home world and venture towards the
Moon. According to the capsule communicator for that phase
of the mission, Michael Collins, soon to become a member of
the historic Apollo 11 crew, Apollo 8 and the first leaving may
in the long term be considered as more important than the
first landing:

I think Apollo 8 was about leaving and Apollo 11 was
about arriving, leaving Earth and arriving at the moon. As
you look back 100 years from now, which is more important, the idea that people left their home planet or the
idea that people arrived at their nearby satellite? I'm not
sure, but I think probably you would say Apollo 8 was of
more significance than Apollo 11, even though today we
regard Apollo 11 as being the showpiece and the zenith
of the Apollo program but, as I say, 100 years from now,
historians may say Apollo 8 is more significant; it's more
significant to leave than it is to arrive.

But what do you say when you leave the Earth for the first time? Collins says:

> I can remember at the time thinking, 'Jeez, there's got to be a better way of saying this,' but we had our technical jargon, and so I said, 'Apollo 8, you're go for TLI.' If, again, 100 years from now you say you've got a situation where a guy with a radio transmitter in his hand is going to tell the first three human beings they can leave the gravitational field of Earth, what is he going to say? He's going to say something like – he's going to invoke Christopher Columbus or a primordial reptile coming up out of the swamps onto dry land for the first time, or he's going to go back through the sweep of history and say something very, very meaningful, and instead he says, 'What? Say what? You're go for TLI? Jesus! I mean, there has to be a better way, don't you think, of saying that?' Yet that was our technical jargon.

Collins said later that he thought there were now three men in the Solar System who would have to be counted apart from all the other billions, three who were in a different place, whose motions obeyed different rules, and whose habitat had to be considered a separate planet.

During the second orbit, at two hours 27 minutes into the mission Mike Collins said, 'You are go for TLI.' Borman responded, 'Roger. We understand; we are go for TLI.' Twenty-three minutes after the on-board computer had completed its calculations, Lovell calmly said, 'Ignition'. The S-IVB had restarted with a long burn over Hawaii that lasted five minutes 19 seconds and boosted speed to the 38,946 km per hour necessary to escape the Earth's gravity. 'You are on your way,' said Chris Kraft, from the last row of consoles in Mission Control, 'you are really on your way.' The anticlimactic observation of the day came when Lovell said, 'Tell Conrad he lost his record'. (During Gemini 11 Pete Conrad and Dick Gordon had set an altitude record of 1,370 km.) After the burn the S-IVB separated and was sent on its way to orbit the Sun.

In Mission Control early in the morning of 24 December, the big centre screen changed to show a scarred and pock-marked map with such labels as Mare Tranquillitatis, Mare Crisium, and many craters with such names as Tsiolkovsky, Grimaldi, and Gilbert. The effect was electrifying, symbolic evidence that man had reached the vicinity of the Moon.

Capcom Gerry Carr spoke to the three astronauts more than 320,000 km away: 'Ten seconds to go. You are GO all the way.' Lovell replied, 'We'll see you on the other side', and Apollo 8 disappeared behind the Moon, the first time in history men had been occulted. For 34 minutes there would be no way of knowing what happened. During that time the 247-second LOI (lunar orbit insertion) burn would take place that would slow down the spacecraft to enable it to go into orbit. If the SPS engine failed, Apollo 8 would whip around the Moon and head back for Earth on a free-return trajectory. During one critical half-minute, if the engine conked out the spacecraft would be sent crashing into the Moon.

'Longest four minutes I ever spent,' said Lovell during the burn, in a comment recorded but not broadcast in real time.

At 69 hours 15 minutes Apollo 8 went into lunar orbit, where-upon Anders said, 'Congratulations, gentlemen, you are at zero-zero.' Said Borman, 'It's not time for congratulations yet. Dig out the flight plan.' Unaware of this conversation, Mission Control buzzed with nervous chatter. Carr began seeking a signal to indicate that the astronauts were indeed in orbit: 'Apollo 8,' he said, 'Apollo 8, Apollo 8.' Then the voice of Jim Lovell came through calmly, 'Go ahead, Houston.'

Mission Control's viewing-room spectators broke into cheers and loud applause. 'What does the old Moon look like?' asked Carr. 'Essentially grey; no colour,' said Lovell, 'like plaster of paris or a sort of greyish beach sand.' The craters all seemed to be rounded off; some of them had cones within them; others had rays. Anders added: 'We are coming up on the craters Colombo and Gutenberg. Very good detail visible. We can see the long, parallel faults of Goclenius and they run through the mare material right into the highland material.' During the second orbit the astronauts captured black-and-white television footage of the Moon (colour would not come until Apollo 10). It proved to be a desolate place indeed, a plate of grey steel spattered by a million bullets. 'It certainly would not appear to be an inviting place to live or work,' Borman said later.

On the third revolution the engine fired for nine seconds to put the spacecraft into a circular orbit where it would stay for sixteen hours (each orbit lasted two hours, as against one and a half hours for Earth orbits). Soon the astronauts were on television again. First, they showed the half Earth across a stark lunar landscape. Then, from the other unfogged window, they tracked the bleak surface of the Moon. 'The vast loneliness is awe-inspiring and it makes you realize just what you have back there on Earth,' said Lovell. The pictures aroused great wonder, with an estimated half a billion people vicariously exploring what no man had ever seen before.

The astronauts got very tired and a little careless during their 20 hours in lunar orbit. Lovell had entered the wrong code into the Command Module's computer, triggering warning alarms, and Anders was overwhelmed with his own tasks. Eventually Borman snapped at Capcom Michael Collins that he was taking an executive decision for his crew to get some rest. 'I'll stay up and keep the spacecraft vertical,' he said, 'and take some automatic pictures.' He had to force Lovell and Anders to pry their eyes away from the windows and sleep.

*

Apollo 8. Lunar Orbit. 24 December 1968. Mission Elapsed Time 75 hours, 47 minutes, 30 seconds.

William Anders:

> We are now approaching lunar sunrise and, for all the people back on Earth, the crew of Apollo 8 has a message that we would like to send to you. In the beginning God created the heaven and the Earth. And the Earth was without form, and void; and darkness was upon the face of the deep. And the Spirit of God moved upon the face of the waters. And God said, Let there be light: and there was light. And God saw the light, that it was good: and God divided the light from the darkness.

Jim Lovell:

> And God called the light Day, and the darkness he called
> Night. And the evening and the morning were the first
> day. And God said, Let there be a firmament in the midst
> of the waters, and let it divide the waters from the waters.
> And God made the firmament, and divided the waters
> which were under the firmament from the waters which
> were above the firmament: and it was so. And God called
> the firmament Heaven. And the evening and the morning
> were the second day.

Frank Borman:

> And God said, Let the waters under the heavens be gath-
> ered together unto one place, and let the dry land appear:
> and it was so. And God called the dry land Earth; and the
> gathering together of the waters called the Seas: and God
> saw that it was good. And from the crew of Apollo 8, we
> close with good night, good luck, a Merry Christmas – and
> God bless all of you, all of you on the good Earth.

*

Borman describes how the astronauts' Christmas message came
about:

> Before the mission Julian Sheer, who was the head of pub-
> lic information for NASA in Washington, called me one
> day. He said, 'You're going to have the largest audience

that's ever listened to or seen a television picture of a human on Christmas Eve; and you've got five or six minutes.' And I said, 'Well, that's great, Julian. What are we doing?' He said, 'Do whatever's appropriate.' That's the only instructions. [...] And to be honest with you, we were so involved in the mission so I just kind of farmed that out to a friend of mine, Si Bourgin, and he came back, well, I guess he consulted with some of his friends and came back with the idea of reading from Genesis. And I discussed it with Bill and Jim, and we had it typed on the flight plan; and I didn't give it any more thought than that.

'At some point in the history of the world,' editorialized the *Washington Post*, 'someone may have read the first ten verses of the Book of Genesis under conditions that gave them greater meaning than they had on Christmas Eve. But it seems unlikely ... This Christmas will always be remembered as the lunar one.' The *New York Times*, which called Apollo 8 'the most fantastic voyage of all times', said on 26 December: 'There was more than narrow religious significance in the emotional high point of their fantastic odyssey.'

As Apollo 8 began its tenth and last orbit, Capcom Ken Mattingly told the astronauts: 'We have reviewed all your systems. You have a GO to TEI' (trans-Earth injection). This time the crew really was in thrall to the SPS engine. It had to ignite in this most apprehensive moment of the mission, else Apollo 8 would be left in lunar orbit, its passengers' lives measured by the length of their oxygen supply. Ignite it did, in a 303-second burn that would affect touchdown in just under 58 hours. Apollo 8 re-entered at 40,000 km per hour and splashed down south of Hawaii two days after Christmas. Aboard the recovery ship, the USS *Yorktown*, President Johnson tried to phone the crew but couldn't be put through for technical reasons. A few minutes later they listened to a tape of him congratulating them.

Mike Collins was however dispirited:

> For me personally, the moment was a conglomeration of emotions and memories. I was a basket case, emotionally wrung out. I had seen this flight evolve in the white room at Downey, in the interminable series of meetings at Houston … into an epic voyage. I had helped it grow. I had two years invested in it – it was my flight. Yet it was not my flight; I was but one of a hundred packed into a noisy room.

The stupendous effect of Apollo 8 was strengthened by colour photographs published after the return. Not only was the technology of going to the Moon brilliantly proven; men began to view the Earth as 'small and blue and beautiful in that eternal silence', as Archibald MacLeish put it, and to realize as never before that their planet was worth working to save. The concept that Earth was itself a kind of spacecraft needing attention to its habitability spread more widely than ever.

During the last week of 1968 the Associated Press re-polled its 1,278 newspaper editors, who overwhelmingly voted Apollo 8 the story of the year. *Time* discarded 'the dissenter' in favour of Borman, Lovell, and Anders; and a friend telegraphed Frank Borman, 'You have bailed out 1968.'

Borman says:

> It's hard for us to fathom now. But the thing that's interest-
> ing about that mission was that, I don't know, maybe half
> a dozen of us sat in Chris Kraft's office one afternoon and
> we went over the flight plan, to try to understand what
> would we do on the whole flight. And I've always thought,
> again, it was an example of NASA's leadership with Kraft
> and their management style that we were able to hammer
> out, in one afternoon, the basic tenets of the mission.

In the USSR, Academician Sedov, still referred to as the 'father
of the Sputnik', told Italian journalists a day after the Apollo 8
splashdown that the Russians had not been competing in a
race to orbit or land on the Moon. Referring to Apollo 8, he
added:

> There does not exist at present a similar project in our
> program. In the near future we will not send a man around
> the Moon, we start from the principle that certain prob-
> lems can be resolved with the use of automatic soundings.
> I believe that in the next ten years vehicles without men
> on board will be the first source of knowledge for the
> examination of celestial bodies less near to us. To this end
> we are perfecting our techniques.

The Central Committee and the USSR Council of Ministers
issued a new decree on 8 January 1969: 'On the Work Plans for
Research of the Moon, Venus, and Mars by Automatic Stations'.

Soon they would say in public that the USSR never wanted to go to the Moon at all; but behind the scenes it was different.

Apollo 8 had been a success and Apollo 9 was planned as an Earth-orbit shakedown of all of Apollo's systems except the landing part. That was the task of Apollo 10, which was planned to go to the Moon and do everything except the landing. But would it be possible, some thought, that Apollo 10 could be the first lunar landing mission.

Early in 1969, George Mueller, NASA's head of the Office of Space Flight, hinted that Apollo 10 *might* make the first lunar landing, but Tom Stafford was not very keen. 'Tom was not so adamant about being first on the Moon,' wrote Cernan in his autobiography. 'He never looked at it that way. He wanted to do what was the best thing to do to have a co-ordinated, planned program.' Stafford told Mueller that if Apollo 10 was retasked to perform the landing, 'this flight crew won't be on it.' Stafford considered that there was just too much work to be done and too many unknowns at that time.

Stafford was right. One of the problems were 'mass concentrations' or 'mascons'. These were regions of higher gravity caused by excess mass, on or just beneath the lunar surface. Areas affected included the impact basins of Imbrium, Serenitatis, Crisium, and Orientale. The effects of these mascons upon satellites was very strong, altering their orbits to the extent of causing them to crash. Mascons were first reported in the journal *Nature* in August 1968 by JPL scientists Paul Muller and William Sjogren, using data from the Lunar Orbiters. By the end of the year it was possible to compile a gravitational map of the Moon's near side. Apollo 8 did not carry a Lunar

Module, so its ability to investigate the mascons was limited. Apollo 10 was the only mission to observe how the Lunar Module's guidance and navigation systems would be affected.

The main problem, though, was that Apollo 10's Lunar Module was overweight. The spacecraft's engineers knew this and LM-4, as it was designated, was meant for either an Earth-orbital or lunar-orbital test flight, not a landing. The next Lunar Module on the production line, LM-5, was built for a landing and was significantly lighter. 'The option, then, was to postpone Apollo 10 for a couple of months until [LM-5] was ready,' wrote Deke Slayton.' 'When you added up what we would gain, as opposed to what we would lose, the decision was pretty easy.'

On 24 March 1969 Neil Armstrong was told that his mission, Apollo 11, would be the first to attempt a lunar landing. He said: 'During the flight of Apollo 8 I had three or four meetings with Deke Slayton about, first, would I take the third one down to the surface and then we had a lot of talks about who might be available and be right to be on that crew, that sort of thing.'

The crew of Apollo 11 – Neil Armstrong, Edwin 'Buzz' Aldrin and Michael Collins – were introduced to the press on 9 January 1969 and immediately the assembled reporters got down to the big question, 'Which of you gentlemen will be the first man to step out onto the lunar surface?'

Over the years, Aldrin has said that he would have preferred to have flown on a later mission. Writing in *Return to Earth* he said: 'I would have preferred to go on a later flight. Not only would there be considerably less public attention, but the flight would have been more complicated, more adventurous, and a far greater test of my abilities than the first landing.' It is clear that for the first few months of 1969 Aldrin believed that he would be the first out of the Lunar Module. He said he had never given it much thought and that he had naturally presumed that he would be first. After all, there were precedents, beginning with Ed White's spacewalk when the commander of the flight stayed in the spacecraft while his partner did the excursions. Aldrin was perhaps right to believe it; NASA's Associate Administrator for Manned Space Flight told several people, including some members of the press, that he would be the first on the Moon.

Word began to filter out, however, that it would be Armstrong. Buzz was angry: Armstrong was a civilian. It would be an insult to the service. He himself was technically still a member of the Air Force, though he had not served for ten years except to maintain his flying hours. So Aldrin approached Armstrong. Aldrin wrote later (although he claimed it was done by his co-author): 'He equivocated a minute or so, then with a coolness I had not known he possessed he said that the decision was quite historical and he didn't want to rule out the possibility of going first.' Armstrong says he cannot remember the conversation. Aldrin talked to his colleagues but for some that was seen as lobbying behind the scenes to be first. Gene Cernan says:

He came flapping into my office at the Manned Spacecraft Center one day like an angry stork, laden with charts and graphs and statistics, arguing what he considered to be obvious – that he, the lunar module pilot, and not

Neil Armstrong, should be the first down the ladder on
Apollo 11. Since I shared an office with Neil Armstrong,
who was away training that day, I found Aldrin's arguments
both offensive and ridiculous. Ever since learning that
Apollo 11 would attempt the first Moon landing, Buzz had
pursued this peculiar effort to sneak his way into history,
and was met at every turn by angry stares and muttered
insults from his fellow astronauts. How Neil put up with
such nonsense for so long before ordering Buzz to stop
making such a fool of himself is beyond me.

That was not the way Buzz saw it. Writing some 40 years later
in his book *Magnificent Desolation,* he said that during training
in early 1969 he recognized that the great responsibility would
fall upon his shoulders. In all the previous Gemini and Apollo
missions, the spacewalks, were taken by the junior officer,
while the commander remained inside the space capsule. As
of February, 1969 that was their plan as well, Aldrin maintained.
On 26 February the *Chicago Daily News* reported that the first
man on the Moon had been selected and it was Aldrin. It read:

> The choice for one of the most momentous events in
> mankind's history falls to the brainy 39-year-old Air Force
> colonel by virtue of his role as Lunar Module commander.
> The disclosure of Aldrin as the choice comes as a surprise
> to many who had speculated that the top commander,
> Neil Armstrong, would be entitled to pull rank and take
> his place in the history books as the first man to set foot
> on a satellite of the earth. But a space agency official said

that the decision is not Armstrong's to make. The flight plan controls the mission and it calls for the lunar module pilot to make the egress … it could be changed but it is not likely to be.

Aldrin said that he hadn't been soliciting the older astronauts' support. He was simply behaving as a competitive Air Force pilot would. 'In truth I didn't really want to be the first person to step on the Moon. I knew the media would never let that person alone.'

On 14 April the speculation came to an end. At a press conference George Low said that the 'plans called for Mr Armstrong to be the first man out after the Moon landing; a few minutes later Colonel Aldrin will follow'. Aldrin later said he believed that the physical layout of the LM dictated that Armstrong went out first. Aldrin, the LM pilot was on the right. But that was not the case. In his biography Armstrong said, 'In my mind the important thing was that we got four aluminium legs safely down on the surface of the moon while we were still inside the craft. But it could technically have been Buzz. Just move before you put the backpacks on.'

Later Chris Kraft explained NASA's thinking. He said that they knew damn well that the first guy on the Moon was going to be a modern-day Charles Lindbergh (the pioneering intercontinental aviator). Neil was calm, quiet and had absolute confidence.

We knew he was the Lindbergh type. He had no ego. The most he ever said about walking on the Moon was that it might have been that he wanted to be the first test pilot to walk upon the Moon. If you would have said to him, you are going to be the most famous human being on Earth for the rest of your life, he would have answered that he didn't want to be the first man on the Moon. On the other hand Aldrin desperately wanted the honor and wasn't quiet in letting it be known. Neil said nothing.

Kraft said that nobody criticized Buzz but that they did not want him to be humanity's ambassador, the man who would be legend. 'The hatch design didn't come into it. That was a rationalization, a solace for Buzz.'

In early March 1969 Apollo 9 on Pad A was ready for launch – delayed by three days because the astronauts caught colds – and the team of 500 engineers and technicians was working twelve or thirteen-hour days. Apollo 10 was ready to be rolled out of the 140-m hangar doors of the VAB for two months of intensive checkout on Pad B. The components of Apollo 11 had already arrived and were undergoing tests in the VAB and in the vacuum chambers of the Manned Spacecraft Operations Building. Apollo 11 would roll out at 12.30, 20 May, one month and 26 days before it was to lift off for the Moon.

A Foreign and Hostile Environment

At last the Lunar Module was ready for testing. Apollo 9 was commanded by James McDivitt. David Scott was the Command Module pilot and rookie Russell Schweickart the Lunar Module pilot. It was to be the most complicated mission so far, testing the docking of the Lunar Module, nicknamed 'Spider', with the Command Module, nicknamed 'Gumdrop', as well as manoeuvres with Spider in free flight. Lift-off was on 3 March 1969, at 11.00, and they soon entered an orbit of 102 by 104 nautical miles. Scott worked the thruster controls as Gumdrop separated from the rest of the spacecraft, turned around, and docked with Spider. It was a major milestone as the procedure was essential to landing on the Moon. About an hour later, the docked spacecraft separated from the Saturn upper stage, using the ejection mechanism for the first time, and about two hours later the crew fired Gumdrop's main engine for five seconds. Throughout the mission Schweickart was sick and felt dizzy, experiencing what is today called space adaptation syndrome, although not much was known about it then. On the second day the crew went through an exhaustive checklist. They fired Gumdrop's propulsion system three more times with a two-hour interval between firings. Two of them were long, almost two minutes and almost five minutes; the third was a shorter burst of about 30 seconds. So far, so good.

McDivitt and Schweickart entered Spider on the third day to check its systems and to fire its so-called descent engine while docked with Gumdrop. Early on the fourth day, Schweickart and Scott began their spacewalks. Schweickart went into the Lunar Module, which was depressurized, and left it by way of the exit hatch that would later be used by the astronauts who landed on the Moon. He stayed around the 'front porch', as the astronauts called it, for about 45 minutes. Scott, meanwhile, had opened the Command Module hatch and clambered partially outside, still hooked to the spacecraft through a life-support umbilical system.

The fifth day was the crucial test for Spider. Schweickart and McDivitt prepared to undock it from Gumdrop. Carefully firing the thrusters, they backed away and rotated so that Scott, in Gumdrop, could see that all the legs were extended, and that there were no obvious problems. They fired the descent engine, moved about 5 km away and for nearly six hours the two astronauts rehearsed procedures. Returning to Gumdrop, they re-docked and jettisoned Spider's descent stage. There was no doubt the spacecraft were ready.

But Apollo 9 also brought home the inherent danger of the Apollo approach. If Jim McDivitt and Rusty Schweickart had been unable to control the LM, Scott might have been able to dock with them manually. But if he couldn't he might have had to abandon them and return home alone, as only the Command Module could re-enter the Earth's atmosphere. He had to be trained for this eventuality. As he later wrote in his book *Two Sides of the Moon*:

> Bringing Apollo home as a one-man show involved my mastering many aspects of all three jobs performed by the crew, Jim's as commander, Rusty's as systems engineer, my own as navigator. The sheer logistics of operating in all three positions, let alone learning the complex procedures this would require, was challenging, to say the least.

A range of disaster scenarios had had to be prepared for. How could Scott rescue the others? What if the two craft docked, but the pressurized tunnel was unusable, or the hatches failed to open? McDivitt and Schweickart would have to exit the LM in their spacesuits and spacewalk back over to the Command Module's side hatch. A major complication was that McDivitt was not scheduled to make a spacewalk and would have to use Schweickart's emergency oxygen supply. 'If he didn't make the spacewalk transfer within 45 minutes,' Scott wrote, 'McDivitt would die.'

Apollo 10, flown by Stafford, Cernan and Young, was different again, because it did go to the Moon – or at least within 14,300 m of it – in its rehearsal of the Apollo 11 mission. Stafford later said that it was not until he was placed into pre-flight quarantine in early May that he realized the magnitude of the mission.

On the evening of the 16th, the night before launch, the prime and backup crews had had dinner with Vice President Spiro Agnew. Late on the following afternoon, driving a little too fast in his rental car back to the Cape Kennedy crew quarters after seeing his family, Cernan was pulled over by a deputy sheriff on Banana River Drive. Cernan had a military driving licence which never expired, which aroused the officer's suspicions. Frustrated, and hoping not to be recognized, when asked for an explanation Cernan said, 'Officer, if I told you, you wouldn't believe me anyway.' The police officer decided that Cernan should go to the police station for further checks. Cernan could see the newspaper headlines, 'Astronaut Arrest Scrubs Moon Mission.'

'Step out of the car, please,' said the police officer but as he reached for his handcuffs, a battered Volkswagen stopped on the other side of the road. In it was Guenter Wendt, in charge of the launch pad during the Apollo 10 launch operation. The German-born Wendt said to Cernan, 'Cheeno! Vaht you doink out here?! You should be gettink ready!' After Wendt explained that Cernan was in the Apollo 10 crew to go to the Moon, the police officer said, 'I've heard a lot of cock-and-bull stories in my life, and if you think I'm gonna believe this one, you're crazy!' He walked away shaking his head, '...get the hell outta here! And go to your Moon!'

'The elevator door rattled closed as we rose up the side of the Saturn 5,' wrote Cernan later, 'higher and higher and we could see clearly through the wide openings of the safety door. Every inch of the way the rocket beside us hummed and vibrated. Glass-like chunks of ice slid away as her cryogenic lifeblood … boiled and bubbled in her guts. *She's alive!*' At length, the elevator stopped and technicians welcomed them. Cernan became aware more than he had ever been that they were heading for the unknown, 'confronting a foreign and hostile environment, where there was no horizon, no up or down, and where speed and time take on new meaning, we not only didn't know the answers … we didn't know the questions!'

Their translunar trajectory was established so precisely that no initial midcourse correction was needed. Stafford says:

We were told that because of the trajectory we would fly, we would not see the Moon until we got there. And that's kind of weird, looking around. You see the Earth go by every twenty minutes, see the Sun go by. Where in the hell is the Moon? And the way the trajectory was and everything, the Moon was eclipsed. Finally, we got, well, a few hours out from the Moon, you could maybe see one little rim of it, hardly a rim. But most of the way to the Moon, we never saw it. Now, we saw it all the way back. Then the Earth started to be eclipsed, and just before we came back to see the Earth, all you could see was a little thin blue line of the Earth. So it was kind of unique and right within a second – BOOM – the Earth goes down. The Earth disappears. There's this big black void. Down below the Earth, when it goes night time, you always see lights and cities and gas fires. There's just all kinds of lights around, and lightning all over. Just a big black void. So we left the Earth. It disappeared. It was quiet. Got turned around, and suddenly – couldn't see anything, and suddenly, about sixty seconds, we were all set, just counting down. Right below us, here comes the Moon, right out in daylight. So it was a real funny feeling there. It really looked weird. And to me, the colour of the Moon in early morning and late at night always looked a little reddish tinge on the top of the mountains. Some people say it's always white and black. I thought it was reddish, with maybe some charcoal grays and tans.

Three days and four hours away from the launch pad, the crew fired the big engine for almost six minutes, inserting them into orbit. Stafford and Cernan entered the Lunar Module, called 'Snoopy', about six hours later to check the systems and then moved back into the Command Module, called 'Charlie Brown', to get some sleep. Then came the climax of the flight. Says Stafford:

We got all squared away and started our manoeuvre to go down to about nine miles above the mountains and do two low passes, check out the landing radar, because if the landing radar doesn't work […] you couldn't land. And it turned out the radar locked on to the lunar surface way in excess of spec, which was good. So as to what we did, we undocked and went way up high above him [John Young in the Command Module] and came down low to get phasing behind him in case we had to abort to come up. So we did that, went down. What always amazed me was the size of the boulders. They were awesome, these big ones, you know, huge things. Some of them are pure

white with black striations up on the side of these gigan-
tic craters. I said, oh, they'd have to be as big as a two- or
three-story building. It's hard to judge distance. Here on
the Earth, even from space, you can still see some roads
and you can see cities. You can kind of judge some distance. No roads up there. No section lines. So anyway,
it turns out those things are bigger than the Astrodome,
those boulders. I mean, they were awesome pieces of mass.

They fired Snoopy's rocket to drop down to within 15,240 metres
of the Sea of Tranquillity. Looking down from Charlie Brown,
Young reported, 'They are ramblin' among the boulders.' As
they made their first pass over the south-western corner of the
Sea of Tranquillity, an excited Cernan called out: 'We is Go,
and we is down among them, Charlie.' Capcom Charlie Duke
responded, 'Roger. I hear you weaving your way up the freeway.'

One of their tasks was to photograph two of the possible
Apollo 11 landing sites. Both lay in the relatively flat Mare
Tranquillitatis region. Attempting to shoot a photograph
every three seconds as Snoopy passed over Site 2, Stafford was
annoyed when the camera jammed. He gave a verbal descrip-
tion and then fired the thrusters to simulate an ascent from
the surface. They were about to fire the four explosive bolts to
separate the ascent stage from the descent stage and begin to
return to Charlie Brown when, suddenly, Snoopy went berserk.
Stafford says:

We were all set to stage off, and I noticed the thrusters
started to fire. I looked down and I could see I had a yaw

rate, but I could tell by the eight ball I wasn't yawing. So I talked to Cernan, and started firing again. We were all buttoned up, and I started troubleshooting, went to the AGS [Abort Guidance System] position and all that, but the first thing you know – BOOM – the whole damned spacecraft started to tumble and tried to rotate like that. And real fast, I just reached over and just blew off the descent stage, because all the thrusters were on the ascent stage, get better torque-to-inertia ratio, because we're heading over towards gimble lock on the main platform. See, it was designed – we had two separate types of platforms. We had a three-gimble platform. We'll never have three-gimble platforms again, you know. So I got it around there, and we got it squared away in about twenty seconds and got ready and lined up, but during that period of time we forgot we were on hot mike, and Apollo 10 became X-rated.

It was pilot error that almost got them killed. Stafford had set a switch controlling the auto guidance system to 'auto', He had mistakenly instructed Snoopy's radar to begin searching for Charlie Brown and the abort guidance system was now causing the lander to spin rapidly. Quickly, he pushed the button to jettison the descent stage and regained control. The incident had lasted three minutes. John Young, listening from Charlie Brown, said, 'I don't know what you guys are doing, but knock it off. You're scaring me!' NASA, and to a certain extent the astronauts, played the incident down. Cernan said he observed the lunar horizon spinning eight times. Just a few more revolutions and it would probably have been unrecoverable and they would have crashed on the surface.

The return rendezvous was flawless. When they ejected the LM ascent stage into solar orbit, it carried the UN flag, a small flag from each state of the Union, empty food packets – and a bag of John Young's faeces from the time when he was alone in the Command Module. 'We joked that Snoopy would have food, water, oxygen, organic material, all the ingredients for the creation of life,' Stafford later wrote with glee. 'Maybe a few billion years from now, some kind of Snoopy monster, distantly related to John Young, will emerge from somewhere in the Solar System …' They fired the Service Module engine to move out of the lunar orbit and back onto a trajectory towards the Earth, using a fast-return flight path that brought them back in 54 hours.

But what if Stafford and Cernan had disobeyed orders and attempted to land? Their LM ascent stage was loaded with the amount of fuel it would have had if it had lifted off from the surface and reached their current altitude, so they just didn't have the fuel. Cernan knew it. He says:

> A lot of people thought about the kind of people we were: 'Don't give those guys an opportunity to land, cause they might!' So the ascent module, the part we lifted off the lunar surface with, was short-fuelled. The fuel tanks weren't full. So had we literally tried to land on the Moon, we couldn't have gotten off.

There was no way the Russians could come close to performing such space missions. None of its lunar spacecraft was anywhere near flightworthy and the Soyuz spacecraft was at best more dangerous than it should be. But despite dampening enthusiasm, a small group of cosmonauts continued to prepare for lunar landings at both the Yuri Gagarin Cosmonaut Training Centre and at the Gromov Flight-Research Institute. In March, veteran cosmonaut Bykovsky was appointed the chief of the lunar department of the cosmonaut detachment. By June it included only eight out of the original group of approximately 25. It was the cosmonauts rather than the politicians or the engineers that forlornly kept the lunar dream alive. Aleksei Leonov said in the spring of 1969:

> The Soviet Union is also making preparations for a manned flight to the Moon like the Apollo program of the United States. The Soviet Union will be able to send men to the Moon this year or in 1970. We are confident that pieces of rocks picked from the surface of the Moon by Soviet cosmonauts will be put on display in the Soviet pavilion during the Japan World Exposition in Osaka in 1970.

But Kamanin wrote in his diary during the Apollo 10 mission of the 'unrestrained lying' by Soviet officials about their intentions with respect to the Moon. He added bitterly, 'We have come to the end to drink the bitter chalice of our failure and be witnesses to the distinguished triumph of the U.S.A. in the conquest of the Moon.'

Throughout the spring and the early summer of 1969, as the world waited for the first manned landing on the Moon, there were many stories in the media speculating that the USSR was planning something spectacular to upstage it. In reality there was little they could do. Even so, von Braun said that it was still possible for the USSR to reach the Moon before the United States if the Apollo 11 mission was delayed, and he strongly believed that the Russians would undertake a piloted lunar flight in the latter part of the year using a giant rocket. But von Braun was out of the loop on this one. The US intelligence community knew the Russians had no chance. In a top-secret CIA 'National Intelligence Estimate' of June they stated that it would be years before they could even attempt a manned lunar flight. Von Braun also talked of the more likely scenario that a Soviet robotic spacecraft would scoop up some soil and bring it back to Earth before Apollo 11 came back with its samples. In fact, the Soviet unmanned lunar sampler mission did indeed have two launch windows to reach the Moon in June and July of that year.

On 4 June a spacecraft was launched from Baikonur in what was a desperate move. It was a gamble but if it paid off, the returned sample of lunar soil, obtained without the need to put a human life at risk, would at least be something to stand against Apollo. But as the third stage of the Proton rocket burned out, a control system failure stopped the next stage from firing. The mission was lost. Instead of going to the Moon, the lunar sampler ended up in the Pacific. The USSR had four remaining lunar scoop spacecraft left and only once chance to beat Apollo 11. Things looked bleak; the Proton rocket had failed on all of its last five missions.

Meanwhile the Apollo prime and backup crews rehearsed and re-rehearsed their movements in simulators and in conditions of weightlessness produced by parabolic flights in a converted Air Force KC-135 tanker and in a training pool. But there was one thing that Apollo 10 didn't do that was causing some consternation. On Apollo 10 the landing radar and the rendezvous radar were never operated at the same time because they were used for two different procedures at two different altitudes. For Apollo 11, however, both radars would be feeding information directly into the computer in the LM. A software review insisted that there should be at least 10 per cent spare capacity in the memory of the on-board computer. This was to be maintained because the memory space might be required in the event of an abort. On Apollo 10, data from the high-altitude radar almost overloaded the computer and set off alarms. Software engineers thought the problem had been dealt with.

But the Russians had not given up completely. The mighty N1, which they hoped would eventually put their cosmonaut onto the Moon, was moved to the launch pad with lift-off set for 3 July, just under two weeks before Apollo 11. Just before midnight, the N1's 30 first-stage rockets burst into life. Lieutenant Menshikov recalled that night:

> We were all looking in the direction of the launch, where the hundred-meter pyramid of the rocket was being readied to be hurled into space. Ignition. The flash of flame from the engines, and the rocket slowly rose on a column of flame. And suddenly, at the place where it had just been, a bright fireball. Not one of us understood anything at first. There was a terrible purple-black mushroom cloud,

so familiar from the pictures from the textbook on weapons of mass destruction. The steppe began to rock and the air began to shake, and all of the soldiers and officers froze.

There was a deathly silence as the onlookers awaited the arrival of the blast wave.

> Only in the trench did I understand the sense of the expression your heart in your mouth. Something quite improbable was being created all around, the steppe was trembling, thundering, rumbling, whistling, gnashing, together in some terrible, seemingly unending cacophony. The trench proved to be so shallow and unreliable that one wanted to burrow into the sand so as not to hear this nightmare, the thick wave from the explosion passed over us, sweeping away and levelling everything. Behind it came hot metal raining down from above.

Pieces of the rocket were thrown 10 km away, and large windows were shattered 40 km away. The 400-kg spherical tank landed on the roof of a building 7 km from the launch pad. Kamanin wrote in his diary:

> Yesterday the second attempt to launch the N1 rocket into space was undertaken. I was convinced that the rocket would not fly, but somewhere in the depth of my soul there glimmered some hope for success. We are desperate for a success, especially now when the Americans intend in a few days to land people on the Moon.

There was only one card left to play – the final chance to launch an unmanned sample return mission. After five straight failures of the Proton rocket, it finally performed well, lifting their last hope off the pad three days before the scheduled launch of Apollo 11. They called it Luna 15. The Soviet media said it was just to study circumlunar space.

'The greatest week in the history of the world'

The responsibility for the Luna 15 mission fell on the shoulders of First Deputy Minister of General Machine Building Georgi Tyulin, a 54-year-old retired artillery general. Tyulin, as chair of Luna 15's State Commission, ran into trouble with the spacecraft after only one day of flight. Controllers detected unusually high temperatures in the propellant tanks that were to be used for take-off from the lunar surface after the collection of the lunar sample. Tyulin assembled all the senior program engineers, including Chief Designer Babakin. After a quick analysis, some participants proposed a seat-of-the-pants method of turning the spacecraft in such a way as to keep the suspect tank in the Sun's shadow at all times.

Luna 15 fired its main engine to enter lunar orbit at 13.00 Moscow Time on 11 July, five days before Apollo 11 took off. Its second orbit correction on 19 July would position it over its landing corridor. If all went to plan, the lander would touch down on the Moon's surface within a couple of hours of the Apollo lander.

At breakfast a few days before the Apollo launch, Thomas Paine, NASA's new Administrator, told Armstrong, Aldrin and Collins that their own safety must govern all their actions, and if anything looked wrong they were to abort the mission. He then made a most surprising and unprecedented statement: if they were forced to abort, they would be immediately recycled and assigned to the next landing attempt. The crew were somewhat puzzled by this and didn't take it seriously. Astronauts assigned to future missions would have had something to say if they had to forgo a landing attempt if Apollo 11 failed.

So the day came. At 04.15 in the Spartan crew quarters in KSC Building 4, Deke Slayton tapped on three doors. 'It's a beautiful day,' he said. The day started like most others, a shower and a shave. A quick medical from Nurse Dee O'Hara who had been the astronauts' personal nurse since Mercury. It was the traditional breakfast before a mission: steak, eggs, orange juice, coffee and toast. Deke Slayton was there, and Bill Anders the backup LM pilot. A NASA artist sketched them as they ate.

They went upstairs to be suited up and, amid the waves and cheers of well-wishers, got into the van to be driven the eight miles to the launch pad, dropping Deke Slayton off along the way. The Sun was rising.

Buzz later said that while Mike and Neil were going through the complicated business of being strapped in and connected to the spacecraft's life-support system, he waited near the elevator on the floor below, as being in the middle seat he was the last to enter the capsule. Alone for fifteen minutes or so in what he described as a 'serene limbo'. He could see there were people and cars lining the beaches and highways and the surf was just beginning to rise out of an azure-blue ocean. 'I could see the massiveness of the Saturn 5 rocket below and the magnificent precision of Apollo above. I savoured the wait and marked the minutes in my mind as something I would always want to remember.'

At this point Michael Collins was thankful that he had flown before, and that the period of waiting atop a rocket was nothing new:

> I am just as tense this time, but the tenseness comes mostly
> from an appreciation of the enormity of our undertaking
> rather than from the unfamiliarity of the situation. I am
> far from certain that we will be able to fly the mission as
> planned. I think we will escape with our skins, or at least
> I will escape with mine, but I wouldn't give better than
> even odds on a successful landing and return. There are
> just too many things that can go wrong.

Fred Haise, the backup astronaut had checked the command-module switch positions and was running through a checklist 417 steps long. Collins at this time had only a few things to do:

> I have plenty of time to think, if not daydream. Here I am,
> a white male, age 38, height 5 feet 11 inches, weight 75 kg,
> salary $17,000 per annum, resident of a Texas suburb, with
> black spot on my roses, state of mind unsettled, about to
> be shot off to the Moon. Yes, to the Moon.

The most important control was on Armstrong's side, alongside his left knee – the abort handle, and it was now powered, so if he were to rotate it 30 degrees counter-clockwise, three solid-fuelled rockets above the capsule on the escape tower would fire and yank it free of the Service Module and everything below it. A large bulky pocket had been added to his left suit leg, and it looked as though he was going to snag the abort handle. Collins pointed this out to Neil, and he grabbed the pocket and pulled it as far over to the inside of his thigh as he could, but it still didn't look secure to either one of them. Collins could imagine the newspaper headlines: 'MOONSHOT FALLS INTO OCEAN. Mistake by crew. Last transmission from Armstrong prior to leaving the pad reportedly was "Oops".'

When all the tests were completed it was time to say goodbye. The astronauts had given gifts to the so-called Pad Führer Guenter Wendt. Collins arranged to have someone bring a small fish from the Banana River the night before the launch. He had the food personnel freeze it and the morning of the launch they mounted the frozen fish onto a wooden plaque bearing the words 'Guenter Wendt/Trophy Trout'. Collins carried the dead frozen fish in the paper bag and presented it to Wendt. Guenter later explained:

> And he [Collins] says, 'Hey, at your house I've never seen a big trophy trout or trophy fish on your wall. You need one.' So now we have a trophy trout. Now there's three things wrong: Illegal size. Not cleaned. And not preserved.

Wendt kept the fish plaque in his freezer for 22 years before getting it preserved. Aldrin gave him a copy of *Good News for Modern Man*, as they were both Presbyterians. Armstrong produced a card that read, 'Space Taxi. Good Between Any Two Planets'.

Armstrong's recollection of the launch is characteristically matter-of-fact. 'The flight started promptly, and I think that was characteristic of all events of the flight. The Saturn gave us one magnificent ride, both in Earth orbit and on a trajectory to the Moon.' Collins says:

> This beast is best felt. Shake, rattle, and roll! We are thrown left and right against our straps in spasmodic little jerks. It is steering like crazy, like a nervous lady driving a wide car down a narrow alley, and I just hope it knows where it's going, because for the first ten seconds we are perilously close to that umbilical tower.

Eleven minutes after lift-off they were in Earth orbit. After one and a half orbits a pre-programmed sequence fired the rocket in the Saturn 5's third stage and sent them on their way to the Moon. After nine hours they were scheduled to make their first midcourse correction, some 91,000 km out. At rocket shutdown, Aldrin recorded their velocity as 10,844 m per second, more than enough to escape from the Earth's gravitational field.

Next Collins had a major task, vital to the success of the mission: separate the Command Module, 'Columbia', from the Saturn third stage, turn around and connect with the Lunar Module, 'Eagle'. Eagle, by now, was exposed; its four enclosing panels had been jettisoned and had drifted away. If the separation and docking did not work they would have to return to Earth. 'Critical as the manoeuvre is, I felt no apprehension about it, and if there was the slightest inkling of concern it disappeared quickly as the entire separation and docking proceeded perfectly to completion,' said Collins later.

Fourteen hours after lift-off, at 22.30 Houston time, they fastened covers over the windows of the slowly rotating command module to get some sleep. Days two and three were devoted to housekeeping chores, a small midcourse velocity correction, and TV transmissions back to Earth. In one news digest from Houston, the astronauts are amused to hear that *Pravda* has referred to Armstrong as 'the czar of the ship'. In the preliminary flight plan Aldrin wasn't scheduled to go to the LM until the next day, when they were in lunar orbit, but he had argued successfully to go earlier. He said he needed to have enough time to make sure its equipment had suffered no damage during the launch. By that time neither Armstrong nor Aldrin had been in the LM simulator for about two weeks.

According to Collins, day 4 of the mission had a decidedly different feel. Instead of nine hours' sleep, he got seven – fitful ones at that. Despite their concentrated effort to conserve their energy, the mental and physical pressure was building. 'I feel that all of us are aware that the honeymoon is over and we are about to lay our little pink bodies on the line,' wrote Collins. 'Our first shock comes as we stop our spinning motion and swing ourselves around so as to bring the Moon into view.' Due to the attitude of the spacecraft, they had not been able to see the Moon for nearly a day, and the change was electrifying. Collins said, 'The Moon I have known all my life, that two-dimensional small yellow disk in the sky, has gone away somewhere, to be replaced by the most awesome sphere I have ever seen.' It was huge, completely filling the window. And it was three-dimensional. The belly of the Moon bulged out towards them in such a pronounced fashion that they felt they could reach out and touch it. To add to the dramatic effect, they could see the stars again as they were now travelling in the shadow of the Moon.

The spacecraft rounded the Moon. They were aware that the Moon was a moving target and that they were racing through the sky just ahead of its leading edge. When they launched, the Moon had been nearly 322,000 km behind where it was now. As they passed behind the Moon, there were only eight minutes before the critical engine burn. On board, things were tense. No one wanted to make a mistake and they were checking and rechecking each step. When the moment finally arrived, the rocket sprang into action and threw them back into their seats. The acceleration was only a fraction of one G but it felt good to them; things were going well. For six minutes they peered at the instrument panel. When the engine shut down Aldrin read out the results: 'Minus one, plus one, plus one.' The accuracy of the overall system was remarkable: out of a total of nearly 1,000 m per second, they had velocity errors of only a few cm per second.

A second burn was to place them in a closer, circular orbit of the Moon, the orbit from which Armstrong and Aldrin would undock and begin their descent in the Eagle. Armstrong and Aldrin started preparing the LM. It was scheduled to take three hours, but because Aldrin had already started the checkout, it was completed a half-hour ahead of time. On the fourth night of the mission they were to sleep in lunar orbit. Although it was not in the flight plan, before covering the windows and dousing the lights, the two hopeful moonwalkers prepared all the equipment and clothing they would need in the morning.

Nobody slept well. Collins remembers the wake-up call: 'Apollo 11, Apollo 11, good morning from the Black Team.' Could they be talking to him? He'd been asleep five hours or so. 'I had a tough time getting to sleep, and now I'm having trouble waking up,' he thought. After breakfast Collins stuffed (his word) Neil and Buzz into the LM along with an armload of equipment. He said:

> Now I have to do the tunnel bit again, closing hatches, installing drogue and probe, and disconnecting the electrical umbilical. I am on the radio constantly now, running through an elaborate series of joint checks with Eagle. I check progress with Buzz: 'I have five minutes and fifteen seconds since we started. Attitude is holding very well.' 'Roger, Mike, just hold it a little bit longer.' 'No sweat, I can hold it all day. Take your sweet time. How's the czar over there? He's so quiet.' Neil chimes in, 'Just hanging on – and punching.' Punching those computer buttons, I guess he means. 'All I can say is, beware the revolution,' and then, getting no answer, I formally bid them goodbye. 'You cats take it easy on the lunar surface …' 'O.K., Mike,' Buzz answers cheerily, and I throw the switch which releases them. With my nose against the window and the movie camera churning away, I watch them go. When they are safely clear of

me, I inform Neil, and he begins a slow pirouette in place, allowing me a look at his outlandish machine and its four extended legs. 'The Eagle has wings,' Neil exults.

Collins made sure all four landing legs were in the correct position.

'I think you've got a fine-looking flying machine there, Eagle, despite the fact you're upside down.' 'Somebody's upside down,' Neil retorts. 'O.K., Eagle. One minute … you guys take care.' Neil answers, 'See you later.' I hope so. When the one minute is up, I fire my thrusters precisely as planned and we begin to separate, checking distances and velocities as we go. This burn is a very small one, just to give Eagle some breathing room. From now on it's up to them, and they will make two separate burns in reaching the lunar surface. The first one will serve to drop Eagle's perilune to 15,000 m. Then, when they reach this spot over the eastern edge of the Sea of Tranquillity, Eagle's descent engine will be fired up for the second and last time, and Eagle will lazily arc over into a twelve-minute computer-controlled descent to some point at which Neil will take over for a manual landing.

Eagle was going lower than Apollo 10, into the unknown. Aldrin says:

Neil and I were harnessed into the LM in a standing position. [Later] at precisely the right moment the engine ignited to begin the twelve-minute powered descent. Strapped in by the system of belts and cables not unlike shock absorbers, neither of us felt the initial motion. We looked quickly at the computer to make sure we were actually functioning as planned. After 26 seconds the engine went to full throttle and the motion became notice-able. Neil watched his instruments while I looked at our primary computer and compared it with our second com-puter, which was part of our abort guidance system.

Gene Kranz was Flight Director in Mission Control, watching everything:

The spacecraft is now behind the Moon, and the control team, the adrenaline, I mean, just really was – no matter how you tried to hide it, the fact is that you were really starting to pump. I mean, the level of preoccupation in these people – and these are kids. The average age of my team was 26 years old. Basically I'm 36; I'm ten years older. I'm the oldest guy on this entire team.

This day, is either going to land, abort or crash. Those are the only three alternatives. So it's really starting to sink in, and I have this feeling I've got to talk to my people. The neat thing about the Mission Control is we have a very private voice loop that is never recorded and never goes anywhere. It's what we call AFD [Assistant Flight Director] Conference Loop. It was put in there for a very specific set of purposes, because we know that any of the common voice loops can be piped into any of the offices at Johnson.

So I called the controllers, told my team, 'Okay, all flight controllers, listen up and go over to AFD Conference.' And all of a sudden, the people in the viewing room are used to hearing all these people talking, and all of a sudden there's nobody talking anymore. But I had to tell these kids how proud I was of the work that they had done, that from this day, from the time that they were born, they were destined to be here and they're destined to do this job, and it's the best team that has ever been assembled, and today, without a doubt, we are going to write the history books and we're going to be the team that takes an American to the Moon, and that whatever happens on this day, whatever decisions they make, whatever decisions as a team we make, I will always be standing with them, no one's ever going to second-guess us. So that's it.

The problems begin. Mission Control can't communicate with Eagle. They have to call Mike Collins to relay data down to them. Going through Kranz's mind is the question of if they've acquired enough data. Is it good enough data so the controllers can make their calls? Are we good? Are we properly configured? The tension mounts at both ends of a 240,000-mile communication loop. Kranz continues:

We move closer now to what we call the 'powered descent go/no go.' This is where it's now time to say are we going down to the lunar surface or not. Now, I have one wave-off opportunity, and only one, and if I wave off on this powered descent, then I have one shot in the next revolution and then the lunar mission's all over. So you don't squander your go/no go's when you've only got one more shot at it.

We lose all data again. So I delay the go/no go with the team for roughly forty seconds, had to get data back briefly, and I make the decision to press on; we're going to go on this one here. So I have my controllers make their go/no go's on the last valid data set that they had. I know it's stale, but the fact is that it's not time to wave off. So, each of the controllers goes through and assesses his systems right on down the line.

We get a go except for one where we get a qualified go, and that's Steve Bales down at the guidance officer console, because he comes on the loop, and he says, 'Flight, we're out on our radial velocity, we're halfway to our abort limits. I don't know what's caused it, but I'm going to keep watching it.' So all of a sudden, boom! We've sure got my attention when you say you're halfway to your abort limits. We didn't know this until after the mission, but the crew had not fully depressed the tunnel between the two spacecrafts. They should have gone down to a vacuum in there, and they weren't. So when they blew the bolts, when they released the latches between the spacecraft, there was a little residual air in there, sort of like popping a cork on a bottle. It gave us velocity separating these two spacecrafts. So now we're moving a little bit faster by the order of fractions of feet per second than we should have at this time. So we don't know it, but this is what's causing the problem.

In the meantime, we've had an electrical problem show up on board the spacecraft, and we've determined that this is a bad meter that we've got for the AC instrumentation. AC, alternating current, is very important on board the spacecraft, because it powers our gyro's landing radar right on down the line. We're now going to be looking at this from the standpoint of the ground so that Buzz won't have to look after it.

All through this time, my mind is really running. Is this enough data to keep going, going, going, going? Because I know what I'm going to do in this role. I'm going to be second-guessed, but that isn't bothering me. We now get to the point where it's time to start engines. We've got telemetry back again. As soon as the engine starts, we lose it again. This is an incredibly important time to have our telemetry because as soon as we get acceleration, we settle our propellants in the tanks, and now we can measure them, but the problem is, we've missed this point. So now we have to go with what we think are the quantities loaded pre-launch. So we're now back to nominals. Instead of having actuals, we've got our nominals in there. So we're in the process of continuing down.

The communication dropouts were a nuisance more than a danger, but the 2011 and 1202 computer alarms could be a showstopper. Armstrong says:

You're always concerned when any kind of alarm comes on, but it wasn't a serious concern because there wasn't anything obviously wrong. The vehicle was flying well, it was going down the trajectory we expected, no abnormalities in anything that we saw, other than the computer said, 'There's a problem, and it's not my fault.' The people here on the ground were right on top of that, and of course, the computer continued in a contrary manner periodically all the way to the surface. But my own feeling was, as long as everything was going well and looked right, the engine was operating right, I had control, and we weren't getting into any unusual attitudes or things that looked like they were out of place, I would be in favour of continuing, no matter what the computer was complaining about.

Kranz says:

So now we're fighting – we've got this new landing area that we're going to be going into, we're fighting the communications, we've got the problem with the communications, and we've got the AC problem that we're now tracking for the crew, and now a new problem creeps into this thing, which is this series of program alarms. There's two types of alarms. These are the exact ones that we blew in the training session on our final training day, twelve-oh-one. Twelve-oh-one is what we call a bail-out type of alarm. It's telling us the computer doesn't have enough time to do all of the jobs that it has to do, and it's now moving into a priority scheme where it's going to fire jets, it's going to do navigation, it's going to provide guidance, but it's basically telling us to do something because it's running out of time to accomplish all the functions it should.

We tell them we're going the alarms, we tell them to accept radar, go on the alarms, you know, radar's good, getting close – you know, we're continuing to work our way down to the surface. Now, fortunately the communications have improved dramatically. Communications are no longer a concern of mine, but they were for about the first six or eight minutes of our descent. But now we're about four minutes off the surface. Communications are just a dream.

Hovering above the lunar surface, Armstrong looked for a landing site. He had taken over manual control at 150 m; the first thing he did was to slow the rate of descent while maintaining his forward speed. There were huge blocks and an extensive boulder field below him. They couldn't land there, they had to press on. Ahead there appeared to be a more open area.

Armstrong says:

We could have tried to land there, and we might have gotten away with it. It was a fairly steep slope and it was covered with very big rocks, and it just wasn't a good place to go. You know, if I'd run out of fuel, why, I would have put down right there, but if I had any choice of a more promising spot, I was going to take it. There were some attractive areas far more level, far less occupied by boulders and things, a half mile ahead or so, so that's where I went.

Duke says:

> When they pitched over to look at the lunar surface, they
> didn't recognize anything and they were going into this
> big boulder field and Neil was flying a trajectory that we'd
> never flown in the simulator. It was something we'd never
> seen. And, you know, we kept trying to figure out […]
> what's going on? You know, he's just whizzing across the
> surface at about 400 feet, and all of a sudden he – the
> thing rears back and he slows it down and then comes
> down. And I'm sitting there, sweating out.

Kranz says:

> Some person – and we've never been able to identify it
> in the voice loop – comes up and says, 'This is just like a
> simulation,' and everybody relaxes. Here you're fighting
> problems that are just unbelievable and you keep working
> your way to the surface, to the surface, to the surface. So
> we get down to the point – and we know it's tough down
> there, because the toe of the footprint is really a boulder
> field, so Armstrong has to pick out a landing site, and he's
> very close to the surface. Instead of moving slowly hori-
> zontal, he's moving very rapidly, and ten and fifteen feet
> per second, I mean, we've never seen anybody flying it
> this way in training.
>   Now [Bob] Carlton calls out 'sixty seconds', and we're
> still not close to the surface yet, and now I'm thinking,
> okay, we've got this last altitude hack from the crew, which

is about 150 feet, which now means that we've got to average roughly about three feet per second rate of descent,
and I see Armstrong's at zero. So I say, 'Boy, he's going
to really have to let the bottom out of this pretty soon. I
crossed myself and said 'Please God.'

Once Armstrong had picked what seemed to be a safe spot it
was a question of lowering the Lunar Module relatively slowly.
They got to within 15 m of the surface and inwardly Armstrong
knew they had done it. Later Gene Kranz said, 'I never dreamed
we would still be flying this close to empty.' When Duke called,
'Thirty seconds', Neil wasn't worried about the fuel. They
landed the simulators with fifteen seconds of fuel left.

Armstrong says:

There was a lot of concern about coming close to running
out of fuel, and I was very cognizant of that. But I did
know that if I could have my speed stabilized and attitude
stabilized, I could fall from a fairly good height, perhaps
maybe forty feet or more in the low lunar gravity, the gear
would absorb that much fall. So I was perhaps probably
less concerned about it than a lot of people watching down
here on Earth.

As it turned out, the touchdown was so gentle that the landing
shock-absorbing springs were hardly compressed.

Then Aldrin said, 'Contact light.' These were the first words
ever said on the lunar surface as a probe from one of the legs registered the ground. A few seconds later Armstrong said, 'Shutdown.'

Kranz says:

Well, what happens, we have a three-foot-long probe stick underneath each of the landing pads. When one of those touches the lunar surface, it turns on a blue light in the cockpit, and when it turns on that blue light, that's lunar contact, their job is to shut the engine down, and they literally fall the last three feet to the surface of the Moon. So you hear the 'lunar contact,' and then you hear, 'ACA [Attitude Control Assembly] out of Detent [out of center position].' They're in the process of shutting down the engine at the time that Carlton says 'Fifteen seconds,' and then you hear Carlton come back almost immediately after that fifteen seconds call and say, 'Engine shutdown,' and the crew is now continuing this process of going through the procedures, shutting down the engine.

Duke says:

Everybody erupted in Mission Control and then his famous lines about, 'Houston. Tranquillity Base here. The Eagle has landed.' And so we made it, you know, and it was really a great release. People cheering and all. I was so excited, I couldn't get out 'Tranquillity Base'. It came out sort of like 'Twangquility'. And so [...] it was, 'Roger, Twangquility Base. We copy you down. We've got a bunch of guys about to turn blue. But we're breathing again.' And I believe that's [...] a true statement. It was spontaneous, but it was true. I mean, we were – I was holding my breath, you know, because we were close.

Kranz:

In the meantime we're just busier than hell [...] we use a cryogenic bottle, super critical helium, to pressurize our descent engine. Again, one of the things you can never test, the heat soak-back from the engine and the surface now is raising the pressure in that bottle very dramatically, and now we're wondering if this damned thing's going to explode and what the hell are we going to do about it. The fortunate thing was that they had designed some relief valves. They had a pressure disc in there. If the pressure got so high, it actually blows the disc and the valve, rather than blowing the bottle up. So we're all sweating this thing out here. We're trying to get everything re-synced for the next lift-off [...] Throughout this whole period of time, except for the instant of hearing the cheering, you never got a chance to really think, 'We've landed on the Moon'.

And so at 02.39 UTC on Monday 21 July 1969, Armstrong opened the hatch, and twelve minutes later began the final part of his descent to the lunar surface. The Remote Control Unit controls on his chest kept him from seeing his feet as he climbed down the nine rungs. On his way he pulled a D-ring to deploy the Modular Equipment Stowage Assembly (MESA) folded against Eagle's side; it deployed and activated the TV camera. An estimated 600 million people were watching.

Afterwards, Armstrong reflected upon his choice of words, 'That's one small step for man. One giant leap for mankind.'

> I thought about it after landing, and because we had a lot of other things to do, it was not something that I really concentrated on but just something that was kind of passing around subliminally or in the background. But it, you know, was a pretty simple statement, talking about stepping off something. Why, it wasn't a very complex thing. It was what it was. I didn't want to be dumb, but it was contrived in a way, and I was guilty of that.

Armstrong said later that there were a lot of things to do, and they had a hard time getting them finished:

> We had very little trouble, much less trouble than expected, on the surface. It was a pleasant operation. Temperatures weren't high. They were very comfortable. The little EMU, the combination of spacesuit and backpack that sustained our life on the surface, operated magnificently. The primary difficulty was just far too little time to do the variety of things we would have liked. We had the problem of the five-year-old boy in a candy store.

Aldrin was jogging to test his manoeuvrability; the exercise gave him an odd sensation. With the bulky suits on they seemed to be moving in slow motion:

I noticed immediately that my inertia seemed much greater. Earth-bound, I would have stopped my run in just one step, but I had to use three of four steps to sort of wind down. My Earth weight, with the big backpack and heavy suit, was 360 pounds. On the Moon I weighed only 60 pounds.

Aldrin said that the view was 'Beautiful, beautiful. Magnificent desolation', adding that he was struck by the contrast between the starkness of the shadows and the desert-like barrenness of the rest of the surface. It ranged from dusty grey to light tan and was unchanging except for one startling sight: the LM was sitting there with its black, silver, and bright yellow-orange thermal coating shining brightly in the otherwise colourless landscape.

I had seen Neil in his suit thousands of times before, but on the Moon the unnatural whiteness of it seemed unusually brilliant. We could also look around and see the Earth, which, though much larger than the Moon the Earth was seeing, seemed small – a beckoning oasis shining far away in the sky.

As they went through their tasks Armstrong had the camera most of the time, and the majority of pictures taken on the Moon that include an astronaut are of Aldrin. It wasn't until they were back on Earth in the Lunar Receiving Laboratory that they realized there were few pictures of Neil. 'My fault perhaps,' said Aldrin, 'but we had never simulated this in our training.'

During a pause in experiments they put up the flag. It took both of them to set it up and it was nearly a disaster. A small telescoping arm was attached to the flagpole to keep the flag extended and perpendicular but it wouldn't fully extend. Thus the flag, which should have been flat, had its own unique permanent wave. Then to their dismay the pole wouldn't go far enough into the lunar surface to support itself in an upright position. After much struggling they finally got it to stay upright, but in a precarious position.

Houston comes on the line saying that the President of the United States would like to talk to them. 'That would be an honor,' says Armstrong.

Neil and Buzz, I am talking to you by telephone from the Oval Office at the White House, and this certainly has to be the most historic telephone call ever made … Because of what you have done, the heavens have become a part of man's world. As you talk to us from the Sea of Tranquillity, it inspires us to redouble our efforts to bring peace and tranquillity to Earth …

Armstrong replies, 'It's a great honor and privilege for us to be here, representing not only the United States but men of peace of all nations, and with interest and a curiosity and a vision for the future.'

They had a pulley system to load the boxes of rocks into the LM's cabin. Back inside they had to pressurize the cabin and begin stowing the rock boxes, film magazines, and anything else they wouldn't need until they were connected once again with Columbia. They removed their boots and the big backpacks, opened the LM hatch, and threw them outside, along with a bagful of empty food packages and urine bags. The exact moment they tossed everything out was measured back on Earth – the seismometer they had deployed was even more sensitive than expected.

Meanwhile, the Luna 15 mission had run into difficulties. Soviet scientists had not anticipated the ruggedness of the lunar surface, and their attempted landing was delayed as a result. Tyutin's State Commission finally commanded Luna 15 to fire its descent engine at 18.47 Moscow Time on 21 July, a little more than two hours prior to the planned lift-off of Armstrong and Aldrin from the Moon. Controllers followed the signals from Luna 15 as it descended. Landing would be in six minutes; but suddenly all data ceased. Later analysis showed that the spacecraft had unexpectedly hit the side of a mountain. TASS announced that Luna 15's research program had been completed and the spacecraft had reached the Moon in the pre-set area.

But even if Luna 15 had worked perfectly and had returned with a soil sample it would have got back to Earth two hours and four minutes after the splashdown of Apollo 11. The race had been over before it was launched.

Before lift-off procedures, Aldrin and Armstrong were scheduled a rest period but they didn't sleep much at all. Lift-off from the Moon, after a stay totalling 21 hours, was exactly on schedule and went exactly as it was supposed to. Aldrin says:

The ascent stage of the LM separated, sending out a shower of brilliant insulation particles which had been

ripped off from the thrust of the ascent engine. There was
no time to sightsee. I was concentrating on the computers,
and Neil was studying the attitude indicator, but I looked
up long enough to see the flag fall over … Three hours
and ten minutes later we were connected once again with
the Columbia.

As Eagle approached, Collins was looking through the docking telescope and saw they were approaching right down the
centre line of the approach path. 'I have 0.7 mile and I got you
at 31 feet per second,' he said. 'For the first time since I was
assigned to this incredible flight, I feel that it is going to happen.
Granted, we are a long way from home, but from here on it
should be all downhill.' After the docking, Aldrin was the first
one through, bearing a big smile. Collins grabbed his head, a
hand on each temple, about to give him a kiss on the forehead,
but then thought better of it and grabbed his hand.

On their way back, the day before splashdown in the Pacific,
they gave a TV broadcast; each had his reflections.

Collins:

The Saturn 5 rocket which put us in orbit is an incredibly
complicated piece of machinery, every piece of which
worked flawlessly. This computer above my head has a
38,000-word vocabulary, each word of which has been
carefully chosen to be of the utmost value to us. […]
Our large rocket engine on the aft end of our service
module, must have performed flawlessly or we would
have been stranded in lunar orbit. The parachutes up

above my head must work perfectly tomorrow or we will plummet into the ocean. We have always had confidence that this equipment will work properly. All this is possible only through the blood, sweat, and tears of a number of people. First, the American workmen who put these pieces of machinery together in the factory. Second, the painstaking work done by various test teams during the assembly and retest after assembly. And finally, the people at the Manned Spacecraft Center, both in management, in mission planning, in flight control, and last but not least, in crew training. This operation is somewhat like the periscope of a submarine. All you see is the three of us, but beneath the surface are thousands and thousands of others, and to all of those, I would like to say, 'Thank you very much.'

Aldrin:

This has been far more than three men on a mission to the Moon; more, still, than the efforts of a government and industry team; more, even, than the efforts of one nation. We feel that this stands as a symbol of the insatiable curiosity of all mankind to explore the unknown. Today I feel we're really fully capable of accepting expanded roles in the exploration of space. In retrospect, we have all been particularly pleased with the call signs that we very laboriously chose for our spacecraft, Columbia and Eagle. We've been pleased with the emblem of our flight, the eagle carrying an olive branch, bringing the universal symbol of peace from the planet Earth to the Moon. Personally, in reflecting on the events of the past several days, a verse from Psalms comes to mind. 'When I consider the heavens, the work of Thy fingers, the Moon and the stars, which Thou hast ordained; What is man that Thou art mindful of him?'

Armstrong:

The responsibility for this flight lies first with history and with the giants of science who have preceded this effort; next with the American people, who have, through their will, indicated their desire; next with four administrations and their Congresses, for implementing that will; and then, with the agency and industry teams that built our spacecraft, the Saturn, the Columbia, the Eagle, and the little EMU, the spacesuit and backpack that was our small spacecraft out on the lunar surface. We would like to give special thanks to all those Americans who built the spacecraft; who did the construction, design, the tests, and put their hearts and all their abilities into those craft. To those people tonight, we give a special thank you, and to all the other people that are listening and watching tonight, God bless you. Good night from Apollo 11.

The evening after the Moon landing someone placed a bouquet of flowers next to the grave of President Kennedy at the Arlington National Cemetery. Attached was a note: 'Mr President. The Eagle has landed.'

They splashed down 900 miles southeast of Hawaii on 24 July and were met by the USS *Hornet*, which was covered with banners saying '*Hornet* + 3'. They were to enter quarantine and on the *Hornet* spoke to President Nixon from behind a small window in the quarantine facility. Nixon said:

> I want you to know that I think I am the luckiest man in the world, and I say this not only because I have the honor to be President of the United States, but particularly because I have the privilege of speaking for so many in welcoming you back to earth.
>
> I can tell you about all the messages we have received in Washington. Over 100 foreign governments, emperors, presidents, prime ministers, and kings, have sent the most warm messages that we have ever received. They represent over 2 billion people on this earth, all of them who have had the opportunity, through television, to see what you have done.

Frank Borman was standing behind Nixon and during his speech and Nixon referred to him a bit later. 'Frank Borman feels you are a little younger by reason of having gone into space. Is that right? Do you feel a little bit younger?'

Armstrong: 'We are younger than Frank Borman.'

Nixon: 'He is over there. Come on over, Frank, so they can see you. Are you going to take that lying down?'

Astronauts: 'It looks like he has aged in the last couple weeks.'

Borman: 'They look a little heavy.'

Concluding, Nixon said:

Well, just let me close off with this one thing: I was thinking, as you know, as you came down, and we knew it was
a success, and it had only been eight days, just a week, a
long week, that this is the greatest week in the history of
the world since the Creation, because as a result of what
happened in this week, the world is bigger, infinitely, and
also, as I am going to find on this trip around the world,
and as Secretary Rogers will find as he covers the other
countries in Asia, as a result of what you have done, the
world has never been closer together before.

On 13 August they took part in parades in New York, Chicago,
and Los Angeles. That same evening in Los Angeles there
was an official state dinner attended by members of Congress,
44 governors, the Chief Justice of the United States, and ambas-
sadors from 83 nations at the Century Plaza Hotel. President
Richard Nixon and Vice President Spiro T. Agnew gave each
astronaut the Presidential Medal of Freedom. It was the start of
a 45-day 'Giant Leap' tour that took them to 25 countries. On
16 September the three spoke before a joint session of Congress
on Capitol Hill. They presented two US flags, one to the House
of Representatives and the other to the Senate, that had been
carried to the surface of the Moon. Life would never be the
same for any of them. Their hardest journey was just beginning.

'Houston, we've had a problem'

The sky was cloudy and rain was falling on 14 November 1969 as the Apollo 12 crew prepared for launch. Half a minute after lift-off a lightning strike opened the main circuit breakers in the spacecraft. Quick action by the crew and Launch Control restored power, and astronauts Charles 'Pete' Conrad Jr, Richard Gordon, and Alan Bean sped into sunlight above the clouds. 'We had everything in the world drop out,' Conrad reported. 'We've had a couple of cardiac arrests down here too,' Launch Control radioed back.

Their destination was the Ocean of Storms, site of many unmanned crashes and soft landings. Four and a half days later, Conrad and Bean entered the Lunar Module 'Intrepid' and separated from Gordon in the Command Module 'Yankee Clipper'. Their landing site was about 2,000 km west of where Apollo 11 had landed, on a surface believed covered by debris splashed out from the crater Copernicus some 400 km away. The exact site was a point where, 31 months before, the unmanned lunar scout Surveyor III had made a precarious automatic landing. The Surveyor site was a natural choice after Apollo 11: it was a geologically different surface, it would demonstrate pinpoint landing precision, and it would offer a chance to bring back metal, electronic, and optical materials that had spent many months in the lunar environment.

Pete Conrad says:

Our first important task was the precision landing near Surveyor III. When we pitched over just before the landing phase, there it was, looking as if we would land practically on target. The targeting data were just about perfect, but I manoeuvred around the crater, landing at a slightly different spot than the one we had planned. In my judgment, the place we had pre-picked was a little too rough. We touched down about 180 m from the Surveyor. They didn't want us to be nearer than 150 m because of the risk that the descent engine might blow dust over the spacecraft.

Al Bean and I made two EVAs, each lasting just under four hours; and we covered the planned traverses as scheduled. We learned things that we could never have found out in a simulation. A simple thing like shovelling soil into a sample bag, for instance, was an entirely new experience. First, you had to handle the shovel differently, stopping it before you would have on Earth, and tilting it to dump the load much more steeply, after which the whole sample would slide off suddenly. And the dust! Dust got into everything. You walked in a pair of little dust clouds kicked up around your feet. We were concerned about getting dust into the working parts of the spacesuits and into the lunar module, so we elected to remain in the suits between our two EVAs. We thought that it would be less risky that way than taking them off and putting them back on again.

On our second EVA we moved on a traverse, picking up samples and describing them and the terrain around them, as well as documenting the specific sites with photography. We rolled a rock into a crater so that scientists back on Earth could see if the seismic experiment was working. (It was sensitive enough to pick up my steps as I walked nearby.) Anyway, we rolled the rock and they got a jiggle or two, indicating that experiment was off and running.

The Surveyor was covered with a coating of fine dust, which looked tan or even brown in the lunar light, instead of the glistening white that it was when it left Earth more than two years earlier. It was decided later that the dust was kicked up by the landing Lunar Module. They cut samples of the aluminium tubing, which seemed more brittle than the same material on Earth, and some electrical cables. Their insulation also seemed dry, hard, and brittle. They broke off a piece of glass, and then unbolted the Surveyor TV camera. Altogether they brought back about 34 kg of rocks, and 7 kg of Surveyor hardware.

While they were busy on the surface, Dick Gordon was busy in lunar orbit. One of the experiments he performed was multispectral photography of the lunar surface, which gave scientists new data about the composition of the Moon. Conrad said:

> After Al and I got back to Yankee Clipper following lunar lift-off and rendezvous, all three of us worked on the photography schedule. We were looking specifically for good coverage of proposed future landing sites, especially Fra Mauro, which was then scheduled for Apollo 13. That's a rough surface, and we wanted to get the highest resolution photos we could so that the crew of the Apollo 13 mission would have the best training information they could get.

As 1970 dawned, so the Apollo program began to diminish. On 14 January NASA announced the cancellation of Apollo 20, which had been scheduled to land near Surveyor 7 in Tycho crater. A day later, writing in the *New York Times*, George Low said that the decision would waste the nation's development.

Nor, for different reasons, would the astronauts of the ill-fated Apollo 13 ever land on the lunar surface. Looking back, says Jim Lovell, commander of the mission, 'I realize I should have been alerted by several omens that occurred in the final stages of the Apollo 13 preparation.' Apollo 13's Command Module pilot, Ken Mattingly, who had been training for the mission for two years, turned out to have no immunity to German measles (a minor disease the backup LM pilot, Charlie Duke, had inadvertently exposed them to). Lovell told Thomas Paine, the NASA Administrator, 'Measles aren't that bad, and if Ken came down with them, it would be on the way home, which is a quiet part of the mission. From my experience as command module pilot on Apollo 8, I know Fred Haise (the LM pilot) and I could bring the spacecraft home alone if we had to.' Paine said no, so backup Jack Swigert stepped in.

The second omen was that during testing they became aware of the possibility that the supercold helium tank in the LM's descent stage might not be properly insulated. To overcome this the flight plan was modified so that during the mission Lovell and Haise entered the LM three hours early to check it out.

Then there was the No. 2 oxygen tank. It had originally been installed in the Service Module of Apollo 10, but was removed to be modified, and became damaged. It was fixed and reinstalled. Tank No. 1 behaved well, but No. 2 dropped to only 92 per cent. An interim discrepancy report was written, and on March 27, two weeks before launch, No. 1 again emptied normally, but No. 2 didn't. It was decided to 'boil off' the remaining oxygen in No. 2 by using the electrical heater inside the tank. Lovell says that with the wisdom of hindsight, he should have said he wanted the tank replacing, 'But the truth is, I went along, and I must share the responsibility with many, many others for the $375 million failure of Apollo 13.'

When Apollo 13, scheduled to be the third lunar landing, was launched at 13.13 Houston time on Saturday 11 April 1970, Lovell had never felt more confident. On his three previous missions, he had already logged 572 hours in space, beginning with Gemini 7, when he and Frank Borman stayed up 14 days. Shortly after the launch, though, things started to go wrong. Lovell says:

The center engine of our second stage of our vehicle shut down two minutes early, probably due to a high vibration which we have a safety feature to shut it off. And for a while there we thought, 'Boy, is there a crisis? Is there a problem with this thing? Do we have enough fuel? Do we have enough power to get into Earth orbit then kick ourselves around to go all the way to the Moon?' Well, very fortunately the folks at Huntsville overbuilt the vehicle. And we did. We had enough fuel. It took us about an extra minute and a half to get into Earth orbit, but we still had enough fuel on the third stage to go all the way to the Moon. We thought that was the crisis.

At 46 hours 43 minutes, Joe Kerwin, the Capcom on duty, said, 'The spacecraft is in real good shape as far as we are concerned. We're bored to tears down here.' At 55 hours 46 minutes, the crew ended a 49-minute TV broadcast with Lovell saying, 'This is the crew of Apollo 13 wishing everybody there a nice evening, and we're just about ready to close out our inspection of Aquarius [the LM] and get back for a pleasant evening in Odyssey [the CM]. Good night.' Listening to the tapes after the mission, Lovell says he sounded mellow and benign, or 'some might say fat, dumb, and happy'. Nine minutes later, all that changed. Gene Kranz says:

And we were down to the final entry, and – the cryogenics, the fuels that we use on board the spacecraft, are oxygen and hydrogen. It's a super dense, super cold liquid at launch at temperatures of -300 to $-400°F$, packed in vacuum tanks. But by the time you're two days into the mission, you've used some of these resources. And these consumables have turned into a very thick, soupy fog or a vapour in the tank. And like fog on Earth, it tends to stratify or develop in layers. So, inside the tanks, we have some fans we turn on to stir up this mixture and make it uniform so we can measure it. Then we use some heaters to raise the pressure for the sleep period. Well, we had asked the crew to do this. In the meantime, the next control team was reporting in for shift hand over, so the noise level in the room was building up; and their flight director, Glynn Lunney (he was the leader of the Black Team, and we used colours to identify those teams), was sitting next to me at the console. He was reading my flight director's log. And we advised the crew that we wanted a cryo stir. Jack Swigert acknowledged our request, and he looked behind him and coming through the tunnel, from the lunar module, was Fred Haise. Sy Liebergot at this time had the responsibilities for the cryo systems, had now switched his attention to the current measurements that he had. And Swigert started the cryo stir.

Liebergot saw the currents increase indicating the stir had started. All of a sudden I get a series of calls from my controllers. My first one is from guidance. It says, 'Flight, we've had a computer restart.' The second controller says, 'Antenna switch.' The third controller says, 'Main bus undervolt.' And then from the spacecraft I hear, 'Hey, Houston, we've had a problem.' And there was a pause for about five seconds and then …

Jim Lovell later said that there was a sharp bang and vibration and that Jack Swigert saw a warning light that accompanied the bang, and said, 'Houston, we've had a problem here.' Lovell:

> Next, the warning lights told us we had lost two of our three fuel cells, which were our prime source of electricity. Our first thoughts were ones of disappointment, since mission rules forbade a lunar landing with only one fuel cell. Houston said, 'Say again, please?' And I say, 'Houston, we have a problem. We have a main B bus undervolt.'

There had been an explosion involving the No. 2 oxygen tank – the one they had had problems with in testing. Kranz continues the story:

> Within Mission Control, literally nothing made sense in those first few seconds because the controllers' data had gone static briefly; and then it – when it was restored, many of the parameters just didn't indicate anything that we had ever seen before. Down in the propulsion area, my controllers all of a sudden saw a lot of jet activity. Jets were firing. We then see Lovell – and this is all happening in seconds – we then see Lovell take control of the spacecraft and fly into an attitude so he can keep communicating with us.
>
> With warning lights blinking on, I checked our situation; the quantity and pressure gages for the two oxygen tanks gave me cause for concern. One tank appeared to be completely empty, and there were indications that the

oxygen in the second tank was rapidly being depleted. Were these just instrument malfunctions? I was soon to find out.

Haise later recalled:

At the time of the explosion, I was in the lunar module. I was still buttoning up and putting away equipment from a TV show we had completed, and really we – subsequently we were going to get ready and go to sleep. I knew it was a real happening, and I knew it was not normal and serious at – just at that instant. I did not necessarily know that it was life-threatening. Obviously I didn't know what had caused it. Within a very short time, though, I had drifted up into the command and service module to my normal position on the right, which encompasses a number of systems – the electrical system, cryogenics, fuel cells, communication, environmental systems – and I was just looking at the array of warning lights. It was confusion in my mind because we had never had a single credible failure that would have caused that number of lights on at one time.

One thing, though, just looking over the instrument panel that became very clear in short order was the fact that the pressure meter, the temperature, and the quantity meter needles for one of the oxygen tanks was down in the bottom of their gauges. These are different sensors, so it was unlikely that this was false. So it effectively told me we had lost one oxygen tank. My emotions at that time went to just a sick feeling in the pit of my stomach, because I knew by Mission Rules, without reference, that that meant the cancellation of the lunar mission.

Lovell says:

The thought crossed our mind that we were in deep trouble. But we never dwelled on it. We never, you know, sort of gave up and said, 'What are – what's going to happen if we don't get back? Where are we going to be?' My thoughts were this: if everything failed and we still had life support in the lunar module but we couldn't get back to the Earth, you know, the heatshield was damaged or we just went past the Earth. I said that, 'We will send back information. We'll keep on operating as long as we can. And then, that's the end of the deal.'

Thirteen minutes after the explosion, I happened to look out of the left-hand window, and saw the final evidence pointing toward potential catastrophe. 'We are venting something out into the – into space,' I reported to Houston. Jack Lousma, the Capcom replied, 'Roger, we copy you venting.' I said, 'It's a gas of some sort.' It was a gas – oxygen – escaping at a high rate from our second, and last, oxygen tank. I am told that some amateur astronomers on top of a building in Houston could actually see the expanding sphere of gas around the spacecraft.

Kranz was very worried:

By this time, Lovell's called down and indicating they're venting something. And we've come to the conclusion that we had some type of an explosion on board the spacecraft; and our job now is to start an orderly evacuation from the command module into the lunar module. At the same time, I'm faced with a series of decisions that are all irreversible. At the time the explosion occurred, we're about 322,000 km from Earth, about 80,000 km from the surface of the Moon. We're entering the phase of the mission – we use the term 'entering the lunar sphere of influence.' And this is where the Moon's gravity is becoming much stronger than the Earth's gravity. And during this period, for a very short time, you have two abort options: one which will take you around the front side of the Moon, and one which will take you all the way around the Moon.

If I would execute what we call a 'direct abort' in the next two hours, we could be home in about 32 hours. But we would have to do two things: we'd have to jettison the lunar module, which I'm thinking of using as a lifeboat, and we'd have to use the main engine. And we still have no clue what happened on board the spacecraft. The other option: we've got to go around the Moon; and it's going to take about five days but I've only got two days of electrical power. So, we're now at the point of making the decision: which path are we going to take? My gut feeling, and that's all I've got, says, 'Don't use the main engine and don't jettison this lunar module.' And that's all I've got is a gut feeling. And it's based, I don't know – in the flight control business, the flight director business, you develop some street smarts. And I think every controller has felt this at one time or another. And I talked briefly to Lunney, and he's got the same feeling.

Then John Aaron said, 'There's no way we're going to make five days with the power in the lunar module. We got to cut it down to at least four days, maybe three and a half.' So, we were now moving ahead. The team split up and moving in several different directions. I had one team working power profiles. I had another group of people that was working navigation techniques. I had a third one that was integrating all the pieces we need. My team picked up the responsibility to figure out […] a way to cut a day off the return trip time.

The knot tightened in my stomach, and all regrets about not landing on the Moon vanished. Now it was strictly a case of survival.

Jim Lovell and his crew had to act swiftly. Lovell takes up the story.

The first thing we did, even before we discovered the oxygen leak, was to try to close the hatch between the CM and the LM. We reacted spontaneously, like submarine crews, closing the hatches to limit the amount of flooding. First Jack and then I tried to lock the reluctant hatch, but the stubborn lid wouldn't stay shut! Exasperated, and realizing that we didn't have a cabin leak, we strapped the hatch to the CM couch. In retrospect, it was a good thing that we kept the tunnel open, because Fred and I would soon have to make a quick trip to the LM in our fight for survival. It is interesting to note that days later, just before we jettisoned the LM, when the hatch had to be closed and locked, Jack did it – easy as pie. That's the kind of flight it was.

The pressure in the No. 1 oxygen tank continued to drift downward; passing 300 psi, now heading toward 200 psi. Months later, after the accident investigation was complete, it was determined that, when No. 2 tank blew up, it either ruptured a line on the No. 1 tank, or caused one of the valves to leak. When the pressure reached 200 psi, it was obvious that we were going to lose all oxygen, which meant that the last fuel cell would also die. At one hour and 29 seconds after the bang, Jack Lousma, then Capcom, said after instructions from Flight Director Glynn Lunney: 'It is slowly going to zero, and we are starting to think about the LM lifeboat.' Swigert replied, 'That's what we have been thinking about too.'

A lot has been written about using the LM as a lifeboat after the CM has become disabled. There are documents to prove that the lifeboat theory was discussed just before the Lunar Orbit Rendezvous mode was chosen in 1962. Other references go back to 1963, but by 1964 a study at the Manned Spacecraft Center concluded: 'The LM [as lifeboat] … was finally dropped, because no single reasonable CSM failure could be identified that would prohibit use of the SPS.' Naturally, I'm glad that view didn't prevail, and I'm thankful that by the time of Apollo 10, the first lunar mission carrying the LM, the LM as a lifeboat was again being discussed. Fred Haise, fortunately, held the reputation as the top astronaut expert on the LM – after spending fourteen months at the Grumman plant on Long Island, where the LM was built.

Apollo 13 was now in survival mode. Everything on the space-craft had to be conserved for re-entry. All the power they had was about the equivalent of a 200 Watt light bulb. Everything else had to be powered down. Another problem occurred: the lithium hydroxide canisters that removed harmful carbon dioxide (CO_2) from the air were in the shut-down Command Module and the canisters in the Lunar Module were running out. The problem was that they were not interchangeable. The canisters in the one wouldn't fit in the other. Using what was in the cabin – air hoses, tape, report covers – a modification was made to remedy the problem.

Water was the real problem. Haise estimated that they would run out of water about five hours before they got back to Earth. But he had an idea. He knew that an engineering test on the Lunar Module showed that it could survive seven or eight hours in space without water cooling, until the guidance system rebelled at the heat.

As they approached the Moon, Mission Control told the crew that they would have to use the LM descent engine a second time; this time a long five-minute burn to speed up their return. The manoeuvre was to take place two hours after round-ing the far side of the Moon. Lovell was busy going through the procedures when he noticed that Swigert and Haise had their cameras out and were busy photographing the lunar surface.

He looked at them incredulously and said, 'If we don't make this next manoeuvre correctly, you won't get your pictures developed!' They said, 'Well, you've been here before and we haven't.'

Haise remembers how cold it got:

We were a little warmer than freezing but not a lot. And that kind of wore on you after a while. We did not have adequate clothing to handle that situation. We did put on […] every pair of underwear we had in the vehicle. Jim Lovell and I wore our lunar boots.

Due to the cold there were fears about what the extensive condensation would do to the controls in the Command Module when it was time to power it up before re-entry. Fortunately it worked. Many believe that was due to the extensive modifications made after the Apollo 1 fire. Prior to re-entry they jettisoned the Lunar Module and the Service Module. They saw that one side of the Service Module had been ripped out by the explosion. The last task was re-entry. Kranz waited nervously:

There isn't any noise in here. You hear the electronics. You hear the hum of the air conditioning occasionally. In those days, we used to smoke a lot. Somebody would only hear the rasp of the Zippo lighter as somebody lights up a cigarette. And you'd drink the final cold coffee and stale soda that's been there. And every eye is on the clock in the wall, counting down to zero. And when it hits zero, I tell Kerwin to, 'Okay, Joe, give them a call.' And we didn't hear from the crew after the first call. And we called again. And we called again.

And we're now a minute since we should've heard from the crew. And for the first time in this mission, there is the first little bit of doubt that's coming into this room that something happened and the crew didn't make it. But in our business, hope's eternal, and trust in the spacecraft and each other is eternal. So, we keep going. And every time we call the crew, it's 'Will you please answer us?' And we were 1 minute and 27 seconds since we should've heard from the crew before we finally get a call. And a downrange aircraft has heard from the crew as they arrive for acquisition of signal. And then almost instantaneously from the aircraft carrier, we get: 'A sonic boom, Iwo Jima. Radar contact, Iwo Jima.' And then we have the 10-by-10 television view. And you see the spacecraft under these three red-and-white parachutes, and the intensity of this emotional release is so great that I think every controller is silently crying. You just hear a 'Whoop!'

Jim Lovell was surprised to discover that the world had been following their progress:

Nobody believes me, but during this six-day odyssey we had no idea what an impression Apollo 13 made on the people of Earth. We never dreamed a billion people were following us on television and radio, and reading about us in banner headlines of every newspaper published. We still missed the point on board the carrier Iwo Jima, which picked us up, because the sailors had been as remote from the media as we were. Only when we reached Honolulu did we comprehend our impact: there we found President Nixon and Dr Paine to meet us, along with my wife Marilyn, Fred's wife Mary (who being pregnant, also had a doctor along just in case), and bachelor Jack's parents, in lieu of his usual airline stewardesses.

Afterwards the builders of the Lunar Module, Grumman Aerospace Corporation sent an invoice for $312,421.24 to North American Rockwell, the builders of the Command Module. It was a fee for 'towing' the crippled ship around the Moon and back to the Earth. The joke invoice included a 20 per cent commercial discount as well as a further 2 per cent discount if Rockwell paid in cash. They declined, saying that their Command Module had ferried their Lunar Module to the Moon on previous occasions free of charge.

Lovell says that since Apollo 13 many people have asked him, 'Did you have suicide pills on board?'

We didn't, and I never heard of such a thing in the eleven years I spent as an astronaut and NASA executive. I did, of course, occasionally think of the possibility that the spacecraft explosion might maroon us in an enormous orbit about the Earth – a sort of perpetual monument to the space program. But Jack Swigert, Fred Haise, and I never talked about that fate during our perilous flight. I guess we were too busy struggling for survival. Survive we

did, but it was close. Our mission was a failure but I like
to think it was a successful failure.

Spurred on by the Apollo 13 setback, the Russians tested their
own version of the lunar lander in Earth orbit in 1970 and 71
but there was nothing to be gained. It took them a while to
realize it.

The Long Goodbye

The Fra Mauro hills stand a couple of hundred kilometres to the east of the Apollo 12 landing site. Apollo 13 was supposed to land there so the site was reassigned to Apollo 14, because scientists gave that area a high priority.

In what is the most remarkable comeback in space history, the commander of the mission was Alan Shepard. He was one of the original Mercury 7 and the first American to fly in space. He had been slated for the first Gemini mission but then contracted Ménière's disease and was grounded.

In 1968, fellow astronaut Tom Stafford told Shepard that an otologist in Los Angeles had developed a cure for Ménière's disease. Shepard flew to Los Angeles, to see Dr William F. House. House proposed to open Shepard's mastoid bone and make a tiny hole to drain excess fluid. The surgery was conducted in early 1969 at St Vincent's Hospital in Los Angeles, where Shepard checked in under the pseudonym of Victor Poulos. It was successful, and he was restored to full flight status a few months later.

Being in charge of astronaut crew selection, Shepard and Slayton put Shepard down to command the next available Moon mission, Apollo 13 in 1970. Under the normal crew rotation procedure Cooper, as the backup commander of Apollo 10, would have been chosen, but Cooper was swept aside. Stuart Roosa, who had not made a spaceflight before was designated the Command Module Pilot and Shepard wanted Jim McDivitt as his Lunar Module Pilot. McDivitt, who had already commanded Apollo 9, was not impressed, saying that Shepard did not have sufficient Apollo training to command Apollo 13. Curiously, another astronaut who hadn't flown in space before, Edgar Mitchell, was then designated as the Lunar Module Pilot.

Slayton had to get approval for flight crew assignments from George Mueller, who rejected the Apollo 13 assignments saying the crew was too inexperienced. Slayton then asked Jim Lovell, the backup commander for Apollo 11, and slated to command Apollo 14, if his crew would be willing to fly Apollo 13 instead. He agreed, and Shepard's inexperienced crew was assigned to the Apollo 14 so that they could get more training. Neither Shepard nor Lovell expected there would be much difference between Apollo 13 and Apollo 14. The failure of Apollo 13 meant that Apollo 14 was delayed until 1971, so that modifications could be made to the spacecraft.

In September, NASA announced that the last two Apollo missions had been cancelled. Visitors to President Nixon's Oval Office noticed that the copy of the Apollo 8 'Earthrise' picture that had been placed on the wall in December the year before had been taken down. The fact was that whenever government cuts were proposed, NASA was at the top of the list.

With Apollo 14 under way, a major problem occurred on 5 February, just prior to the final descent at Fra Mauro, when Shepard and Mitchell were in the LM, 'Antares'. An abort command was received by the Lunar Module's guidance computer. Had the abort command been initiated, it would have separated the ascent stage from the descent stage, ending the landing altogether. As it was, the descent had to be delayed as Mission Control investigated. They concluded that the problem was that the abort switch itself was faulty. A computer program was written and tested within two hours by the operations team and inserted manually into the computer by Mitchell. According to Shepard, the targeting data for the Apollo 14 landing site were every bit as good as the data for Apollo 12; nevertheless, they had to fly around for a little while. As for Apollo 12, this was because the landing site was rougher, on direct observation, than the photos had indicated.

On the first moonwalk Shepard and Mitchell set up the solar-wind experiment and the flag, and deployed the surface science package. The latter had two new experiments. One was called the 'thumper': Ed Mitchell set up an array of geophones, and then walked out along a planned survey line with a device that could be placed against the surface and fired, to create a local impact of known size. Thirteen of the 21 charges went off, with good results. The other experiment was a grenade launcher, with four grenades to be fired off by radio command sometime after they had left the Moon. They were designed to impact at different distances from the launcher, to get a pattern of seismic response to the impact explosions. While the moon-walkers were performing their tasks, Stuart Roosa was obtaining photographic coverage of the proposed site for the Apollo 15 mission, near the Descartes crater.

During their first moonwalk, Shepard and Mitchell worked on the surface for four hours and 50 minutes. For their second EVA they used the MET – Modularized Equipment Transporter, although they called it the lunar rickshaw – to carry tools, cameras, and samples so they could work more effectively and bring back a larger quantity of samples. According to Shepard:

> Our planned traverse was to take us from Antares more or less due east to the rim of Cone crater. That traverse had been chosen because scientists wanted samples and rocks from the crater's rim. The theory is that the oldest rocks from deep under the Moon's surface were thrown up and out of the crater by the impact, and that the ones from the extreme depth of the crater were to be found on the rim.
>
> On our way to the crater, one of the first things Ed did was to take a magnetometer reading at the first designated site. When he read the numbers over the air, there was some excitement back at Houston because the readings were about triple the values gotten on Apollo 12. They were also higher than the values Stu was reading in the Kitty Hawk [the Command Module], and so it seemed that the Moon's magnetic field varied spatially. Our first sampling began a little further on, in a rock field with boulders about two or three feet along the major dimension. These were located in the centers of a group of three craters, each about sixty feet across. Like the bulk

of the samples brought back, these were documented samples. That means photographing the soil or rocks, describing them and their position over the voice link to Mission Control, and then putting the sample in a numbered bag, identifying the bag at the same time on the voice hookup.

The mapped traverse was to take them nearly directly to the rim of Cone crater, a feature about 300 m in diameter. As they approached, the boulders got larger, up to 1.5 m in size. And at this time, the going started to get rougher:

> The terrain became more steep as we approached the rim, and the increased grade accentuated the difficulty of walking in soft dust. Another problem was that the ruggedness and unevenness of the terrain made it very hard to navigate by landmarks. Ed and I had difficulty in agreeing on the way to Cone, just how far we had travelled, and where we were.

As they moved towards Cone, into terrain that had almost continuous undulations, and very small flat areas, the surface began to slope upward even more steeply, and they felt they were starting the last climb to the rim of Cone. They passed a rock which had a lot of glass in it, and reported to Houston that it was too big to pick up. As they continued, they altered their suit cooling rate to match their increased work output as they climbed. For a while, they picked up the cart and carried it, preferring to move that way because it was a little faster. And then came what had to be one of the most frustrating experiences on the traverse. Shepard tells the story:

> We thought we were nearing the rim of Cone, only to find we were at another and much smaller crater still some distance from Cone. At that point, I radioed Houston that our positions were doubtful, and that there was probably quite a way to go yet to reach Cone. About then, there was a general concurrence that maybe that was about as far as we should go, even though Ed protested that we really ought to press on and look into Cone crater. But in the end, we stopped our traverse short of the lip and turned for the walk back to Antares.

Later estimates indicated they were perhaps only 10 m or so below the rim of the crater, and yet they were not able to define it in that undulating and rough country.

> We stopped at Weird crater, for more sampling and some panoramic photography, and then continued the return traverse. At the Triplet craters, more than three-quarters of the way back to Antares, we stopped again. Ed's job there was to drive some core tubes; I was to dig a trench to check the stratification of the surface. But the core material was granular and slipped out of the tube every time Ed lifted it clear of the surface. I wasn't having any better luck with my trenching, because the side walls kept collapsing. I did get enough of a trench dug so that I could observe some stratification of the surface materials, seeing their

colour shift into the darker browns and near blacks, and then into a surprisingly light-coloured layer underneath the darkest one.

By the time they got back to Antares they had covered a distance of about 3 km and collected many samples during four and a half hours on the surface in the second EVA. 'I also threw a makeshift javelin, and hit a couple of golf shots,' said Shepard. 'That was our mission. Our return was routine, our landing on target, and our homecoming as joyous as those before.' He continues:

I look back now on the flights carrying Pete's crew and my crew as the real pioneering explorations of the Moon. Neil, Buzz, and Mike in Apollo 11 proved that man could get to the Moon and do useful scientific work, once he was there. Our two flights – Apollo 12 and 14 – proved that scientists could select a target area and define a series of objectives, and that man could get there with precision and carry out the objectives with relative ease and a very high degree of success. And both of our flights, as did earlier and later missions, pointed up the advantage of manned space exploration. We all were able to make minor corrections or major changes at times when they were needed, sometimes for better efficiency, and sometimes to save the mission.

Apollo 12 and 14 were the transition missions. After the cart towed by the moonwalkers of Apollo 14 came the Lunar Rover:

a wheeled vehicle to extend greatly the distance of the traverse and the quantity of samples that could be carried back to the Lunar Module.

In July 1971 Apollo 15 hit its target precisely, a large amphitheatre girded by mountains and a deep canyon on the eastern edge of a vast plain. Later, Commander David Scott said he would never forget the Command Module, 'Endeavour', hurtling through the Moon's strange night-time. Above were the stars, below lay the Moon's far side, an arc of impenetrable darkness. As the moment of sunrise approached, barely discernible streamers of light – actually the glowing gases of the solar corona around the Sun – played above the Moon's horizon. Finally, the Sun exploded into his view. In less than a second its harsh light flooded Endeavour and dazzled his eyes. The early lunar morning stretched into the distance. Long angular shadows accentuated every hill and crater. As the Sun rose higher, the moonscape turned the colour of gunmetal.

Dave Scott and Jim Irwin spent 67 hours on the Moon, landing in the bright morning of the 710-hour lunar day. Opening the top hatch, Scott made a preliminary survey, looking out on a world he described as still being in the epoch of its creation. Craters left by 'recent' meteorites millions of years ago stood out startlingly white against the soft beige of the gently undulating terrain. The Lunar Module, 'Falcon', had landed on the edge of the so-called Sea of Rains – Mare Imbrium – which stretches across a swathe of the Moon for over 1,000 km. To the south of Falcon, a 3,300-m ridge rose over the plain. To the east was an even higher summit. To the west was the Hadley

Rille, snaking across the landscape and 300 metres deep. To the north-east was a great mountain towering 5,000 metres above them. 'Their majesty overwhelmed me,' said Scott.

Scott was a space veteran on his third flight but he was also a new breed of astronaut. Eight years of training in lunar geology made him aware of intriguing details in the landscape and the rocks – a dark line, like a bathtub ring, smudged the base of the mountains. Was it left by the subsiding lake of lava that once filled the immense cavity of Palus Putredinis on the fringes of Mare Imbrium billions of years ago?

On the surface, Scott found the one-sixth Earth gravity more enjoyable than weightlessness, in that it retained the same sense of buoyancy but with a reassuringly fixed sense of up and down. He felt like an intruder, he said, in an eternal wilderness. The flowing moonscape reminded him of the Earth's uplands after a covering of snow. Most of the scattered rocks shared the same greyness as the dust. However, he found two that were jet-black, two that were pastel green, several with sparkling crystals, some coated with glass and one that was white. No wind blew, no sound echoed, only shadows moved.

At first, Dave Scott and Jim Irwin experienced a troubling deception in perspective. There were no trees, clouds or haze to determine whether an object were far or near. Each of the three spacewalks was due to last seven hours and they dug and drilled, gathered rocks, took photographs. Back in Falcon between excursions, it took them two hours to remove their suits and do housekeeping chores. For the first twenty minutes or so they were aware of a smell like that of gunpowder, before the air filtration system purified the air. The moondust stuck to everything. To sleep, they put shades over the windows before settling into hammocks.

By the third moonwalk they felt at home. Using the Lunar Rover on its first mission they ventured over the horizon – the first astronauts ever to do so. In case of problems with the navigation system on the rover, Scott had made a small cardboard compass that used the Sun's position as a reference since its position didn't change much during their brief stay. Although shrivelled in the savage lunar sunlight and covered with moondust, it would give him the bearings back to Falcon if he were to need them. On their way back the astronauts even dared to take a short cut, the rover bouncing between undulations and crater walls that obscured their view of the Lunar Module for long minutes.

At leaving the Moon they felt a sense of impending loss. They would never return to the plains of Hadley. Clutching the ladder, Scott raised his eyes from the now-familiar moonscape and saw the vivid blue sphere of the Earth.

On the descent stage of the lander, as with all Apollo landers, a plaque of aluminium portrayed the two hemispheres of Earth, as well as giving the name of the spacecraft, the date of the mission and the roster of the crew. The crew of Apollo 15 left behind a falcon's feather and a four-leaf clover. In a little hollow in the moondust they placed a stylized figure of a man in a spacesuit alongside a metal plaque bearing the names of the fourteen Russian and American spacemen who have given their lives so that man may explore the cosmos. Alongside, Scott lays a single book, the Bible.

In April 1972 Charles Duke, the astronaut who communicated with Armstrong and Aldrin as they landed on the

Moon, got his chance to make his own landing with Apollo 16 – although when they reached lunar orbit, a malfunction in the Service Module controlling the angle of the booster nozzle nearly caused the cancellation of the lunar landing.

One of the Command Module pilot Ken Mattingly's tasks was to fire the Service Propulsion System engine to adjust his orbit from an ellipse into a near-circular one to enable a three-day program of lunar science with the instruments in the Service Module. Mattingly went through the procedure and suddenly the CSM, named 'Casper', shuddered.

'I have an unstable yaw gimbal No. 2,' he radioed to Young and Duke.

'Oh, boy,' replied Young, who knew it could cancel the landing.

In consultation with Mission Control, four hours later, it was decided to use a secondary system to control the burn. What no one except Mattingly knew was that all of the rocket motor's control signals – main and backup – used the same cable. After Apollo 16 returned to Earth, it was said by Mission Controllers that had they known about the cable, they would never have allowed the landing.

Another problem with communications meant that Young and Duke had to start their descent with the windows pointed out to space and just depend on the landing radar to update them on their altitude. Duke says:

So at 7,000 feet, the guidance program manoeuvred the vehicle to windows forward down, and I saw the lunar surface close up for the very first time. It looked exactly like the mock-up. 'John, there it is!' you know, 'There's Gator. There's Lone Star and North Ray Crater.'

John Young descended the ladder of the LM and became the ninth man on the Moon. 'There you are, our mysterious and unknown Descartes highland plains,' he said. 'Apollo 16 is gonna change your image!'

Looking back, Young describes the mission:

We landed in the Descartes highlands of the Moon – a valley 8 to 10 miles across, and the objective was to explore to the south to a place we called Stone Mountain and then to the north, 3 or 4 miles, to a place called North Ray Crater, which was at the base of the Smoky Mountains. With the rover, you could do that. It took us 40–50 minutes to drive down south; and I was the navigator. We had trained, so I was the navigator; and John was the driver of the rover.

We landed within a couple of hundred meters of where we thought we were going to land. So we [...] basically recognized the major landing spots. And I remember as John started off, I said, 'Okay, John. Steer 120 degrees for 1.2 kilometres, and then turn left to 090 degrees and go another 2 kilometres' or whatever it was. And so, that's the way we navigated. The lunar rover had a little directional gyro. There was no magnetic field on the Moon, so a magnetic compass wouldn't work. So we had a little gyroscope that was mounted in the instrument panel of the rover. We had a little odometer on the wheel that counted out in kilometres, and so that was our distance. And so, that's how we navigated up on the lunar surface. We'd start out one direction and we'd make a big loop and come back to the lunar module 6–7 hours later. That was the plan. And, you never really worried about getting lost up there because everywhere you drove, you left your tracks. And so, if you really were unsure of your position, it was easy just to turn around and follow your tracks back.

We kept jogging and jogging, and the rock kept getting bigger and bigger and bigger. And we were going slightly downhill, that we didn't sense at first, and so we get down to this thing and we called it 'House Rock'. You know, it must've been 90 feet across and 45 feet tall. It was humongous. And we walked around to the front side or the east side, which was in the sunlight, and, you know, it was towering over us and John and I hit it with a hammer, and a chunk came off, and we were able to collect a piece of House Rock. But – then we had to hike back. It was uphill, and it was a struggle getting back up.

Before he left, Charlie Duke left on the surface a picture of his family. A message on the back reads: 'This is the family of Astronaut Duke from Planet Earth. Landed on the Moon, April 1972.' Underneath are the signatures of his wife and children.

On 6 December 1972, Gene Cernan, Harrison Schmitt and Ron Evans were in the Command Module 'America', on top of a fuelled Saturn 5 for the first night-time launch in US spaceflight history. The launch time was dictated by the angle of the Sun at the landing site in the Taurus-Littrow highlands when they arrived – the shadows cast had to be long enough to show sufficient relief to allow a safe landing.

Gene Cernan was making his third trip into space and his second Apollo mission, having also flown on Apollo 10. This time he was to take part in what would be the final Apollo Moon landing, along with geologist Harrison Schmitt. Originally Joe Engle was due to accompany Cernan but when it was realized that Apollo 17 would be the last such mission, NASA decided to replace him with a professional geologist who had been training for the now-cancelled Apollo 18.

With the exception of Apollo 14, the launch of which was delayed 40 minutes due to the weather conditions, every manned Apollo launch up to this point had lifted off exactly on schedule. Now, however, Apollo 17 was delayed two hours and 40 minutes, until 12:33 a.m., because of the failure of an automatic countdown sequencer in the ground equipment. Gene Cernan said later that he kept his hand very tightly on the abort switch, 'because you never know'.

Cernan recounts the experience of landing on the Moon:

Our valley where we were to land in was surrounded by mountains on three sides that are higher than the Grand Canyon is deep, to give you some idea. So at 7,000 feet we were down among them. I mean the mountains rose above us on both sides. The valley was only 20 miles long and about five miles wide. We had good photography. So [we] practiced this 100, 500, I don't know how many times. So what I was looking at I'd seen before basically, because of the simulation and the pictures. So I knew we were in the right spot. At 7,000 feet as the craters and rocks and the boulders and so forth began to appear I could begin to pick up my landing site.

We had a particular target point, but it was only as good as we expected it to be. But when I got closer and I could see, then I could what we called re-designate where we were going to land. As I say, all the way down the engine is firing, you're in a suit, it's noisy, it's vibrating, people are talking to you from both ends, needles are going left and right. You know you don't have much fuel. So you got to get down quickly. But you can't get down too quick, you got to have your rate of descent under control.

You get down to 200 feet and you're going to land or crash because if something happens to the descent engine at that point in time you can't react quick enough to stage the two vehicles, fire the ascent engine, and get out of there. So when you're down below 200 feet you're going to land. I mean I wasn't going to go all that way a second time and not land. Fortunately everything worked well for us. The landing radar, all the equipment, everything worked fine. You're coming down pretty fast through 200 feet. I don't remember exactly, but somewhere around 30, 35 feet per second, which is pretty fast, you've got to slow down from that point on so that you touch down at one or two feet per second. You get to about 80 feet, and you start blowing dust all over the place. By that time you now know where you're going to land, the dust keeps you from really seeing much of anything, because it just scatters horizontally in all directions. You effectively take what you got.

'It could have been two seconds, ten seconds, a minute or two, I don't know. But after we all got our breath and realized 'hey, we are there', that's when I told Houston, 'Houston, the Challenger has landed.' There have been people who want to believe in the fantasy or the conspiracy, whatever, that it was all done in Hollywood, we never really walked on the Moon. Well, if they want to have missed one of the greatest adventures in the history of mankind, that's their choice. But once my footsteps were on the surface of the Moon, nobody, but nobody, could ever [...] take those footsteps away from me. Like my daughter's initials I put into the Moon during that three days we were there. Someone said, 'How long will they be there?' I said, 'Forever, however long forever is.' I'm not sure we, any of us, understand that.

As Cernan and Schmitt ranged over the Moon, back in Houston Cernan's daughter Tracy was interviewed on *The Today Show*. She was asked if her father was bringing her back something special. 'I can't tell you,' she said. The interviewer persisted. She gave in. 'He's going to bring me back a Moonbeam.'

Soon it was time to leave the Moon for the final time. With a waning gibbous Earth in the black sky above him, Cernan said goodbye.

When I climbed up the ladder for that last step, and I looked down, and there was my final footsteps on the surface, and I knew I wasn't coming back this way again, somebody would – and somebody will – but I knew I was not going to come back this way. I looked over my shoulder because the Earth was on top of the mountains in the southwestern sky. Never moved for the whole three days we were there. People kept saying, 'What are you going to say, what are going to be the last words on the Moon?' I never even thought about them until I was […] basically crawling up the ladder.

'As I take man's last step from the surface, back home for some time to come – but we believe not too long in the future – I'd like to just say what I believe history will record – that America's challenge of today has forged man's destiny of tomorrow. And as we leave the Moon at Taurus-Littrow, we leave as we came and, God willing, as we shall return with peace and hope for all mankind.'

On the leg of the descent stage of 'Challenger', the Lunar Module, is a plaque that reads: 'Here Man completed his first explorations of the Moon. December 1972 AD. May the spirit of peace in which he came be reflected in the lives of all mankind.' Nearby, next to the abandoned Lunar Rover, drawn in the lunar soil, are the initials, TDC, standing for Tracy Dawn Cernan, who was waiting to welcome her father home. Apollo 17 splashed down on 19 December 1972. Within days of its return, US National Public Radio carried an interview with a farmer from Ohio who said, 'I don't think they went to the Moon.' Since then no one has left low Earth orbit.

Apollo 17 brought back 11 kg of lunar samples. One of them is particularly cherished by geologists. It is a 1-m-long tube of regolith taken from a deep drilling. It is kept as a reserve sample and has never been opened.

The Melancholy of
All Things Done

In the final summary, 29 astronauts directly took part in the Apollo Moon program. Between December 1968 and December 1972 24 of them went to the Moon (three twice) and twelve walked upon it. When they walked Charlie Duke was the youngest at 36, and Alan Shepard was the oldest at 47. Of the twelve only four are still living at the time of writing; Buzz Aldrin is the eldest at 89. Since then plans to return to the Moon have come and gone and none of them have lasted very long. Neil Armstrong used to draw a diagram at some of his lectures. It consisted of four circles, inside which he wrote 'Leadership', 'Threat', 'Good Economy' and 'World Peace'. He would then say that when you get all of them in conjunction you can do something like Apollo. What he was asking was, are these the only conditions when mankind is motivated to do things like go to the Moon, or develop nuclear energy? Perhaps they are. Unless there is a more sustained political will and public support there will come a time when there are no moonwalkers among us.

After returning from the Moon Neil Armstrong worked for NASA in the Office of Advanced Research and Technology but his work was inevitably interrupted by public appearances and it seems NASA did little to help him manage his commitments. He resigned from NASA in August 1971 and went to the University of Cincinnati, 'I just wanted to be a university professor,' he said. But the first man to walk on the Moon couldn't be 'just' a university professor. In 1979 he resigned to be national spokesman for Chrysler automobiles and to sit on various corporate boards, and try to maintain a lower profile. Not everyone thought Armstrong should be so withdrawn, especially at a time when NASA was developing the Space Shuttle and had turned its back on going any further than low Earth orbit.

Jim Lovell said to him that he should not be so Lindbergh-like. Lindbergh, he said, flew across the Atlantic using private money, so he had a right to be withdrawn. Armstrong's trip was paid for by taxpayer's money, and it was they, as well as his own desire, that had put him in the position of being the most famous man on Earth. There must therefore be a certain amount of return due to them. Armstrong's response was that he would be harassed all the time if he wasn't reclusive. In 1986 he sat on the inquiry panel into the Challenger Space Shuttle accident, but during its investigations things were not good in Armstrong's private life.

In 1990 his parents died and his wife Jan left him. Later she said she cried for three years before she left. A year later he suffered a heart attack. In 1994 Jan divorced him, despite Armstrong begging her not to. 'The man needed help,' she said.

'I couldn't help him. He really didn't want me helping him.' During those difficult years, mutual friends introduced him to Carol Knight, and they married in 1994. In the subsequent years he seemed much happier and met a lot of people and gave a lot of talks, albeit with long silences in between, and rarely on a grand scale except for Apollo anniversaries. To the public he was a great American hero *in absentia*.

But what was the alternative? Chat shows, red carpets, holidays with celebrities at ski resorts, dancing with the stars, being Captain America?

Armstrong underwent bypass surgery on 7 August 2012 to relieve blocked coronary arteries. Although he was reported to be recovering well, he developed complications in the hospital and died on 25 August in Cincinnati, Ohio, aged 82. Aldrin, said that he was 'deeply saddened by the passing. I know I am joined by millions of others in mourning the passing of a true American hero and the best pilot I ever knew. I had truly hoped that on July 20th, 2019, Neil, Mike and I would be standing together to commemorate the 50th anniversary of our Moon landing … Regrettably, this is not to be.'

At the memorial service on 13 September at Washington National Cathedral, Gene Cernan, the last man to walk on the Moon, said:

Fate looked down kindly on us when she chose Neil to be the first to venture to another world, to have the opportunity to look back from space at the beauty of our own. It could have been another, but it wasn't, and it wasn't for a reason. No one, no one but no one, could have accepted the responsibility of his remarkable accomplishment with more dignity and grace than Neil Armstrong.

Hearing those words from the large dais near the pulpit was Michael Collins, who read a prayer. Sitting in the front row in the Cathedral's wing was Buzz Aldrin. On his left were Annie and John Glenn. On his right was Christina Korp, manager of Buzz Aldrin Enterprises. The following day, Armstrong's cremated remains were scattered in the Atlantic Ocean during a burial at sea ceremony aboard the USS *Philippine Sea*.

Buzz Aldrin also struggled with life after Apollo. His son said that everyone wanted Aldrin to play the hero and that they created a role for him that he didn't want. Before his trip to the Moon there had always been another mission. But this time there was no next mission to focus on, and Aldrin felt it deeply. Like his crewmates, he was not yet 40 years old, with a future defined by others and by the past.

In 1972 he re-joined the Air Force, for a while, as commandant of the Test Pilot School at Edwards Air Force base. His friends were worried about this move. Tom Stafford told him, 'You're a good fighter pilot, you shot down two and a half MiGs in Korea … but remember, you're still not a test pilot. If you are going to go, go with some humility, and listen. Don't talk! Don't be in transmit mode.' Aldrin lasted six months and took retirement on medical grounds.

His wife said no one offered him a job. Consequently the Aldrins (like the families of the other astronauts) never had any money. He was forced to sell his house and slipped into depression and alcoholism. He and Joan divorced.

In his books Aldrin has been brutally honest about his drinking and mental health issues and the depths of despair he reached. His is perhaps the best account of the other side of travelling to the Moon, the counter-journey with the human spirit in pain. It was his most difficult mission and he was unafraid to provide a full briefing. It was said that perhaps no one ever truly knew Armstrong – all was deeply hidden. With Aldrin one gets the impression of a man who has tamed an inner turbulence, and found his place in the world.

In 1988 he married Lois Driggs Cannon, who got him organized and acted as his manager in what he later called the 'business of Buzz'. He formally changed his name to Buzz and began to be more commercial. After all, Armstrong was a recluse so he had a free market. But eventually his marriage collapsed and he filed for divorce on 15 June 2011, in Los Angeles. The ensuing court case attracted worldwide publicity, and Lois won half of Aldrin's money and substantial living expenses. These days Aldrin travels the world extensively, moving between conferences and endorsements with the energy of a man half his age. He advocates a manned mission to Mars, and often wears a T-shirt with 'Get Your Ass To Mars' on it, a quote from the science fiction film *Total Recall*.

The final member of the crew of Apollo 11, Command Module Pilot Michael Collins, has had the quietest life of the three. Before the mission he had already said he would be leaving the astronaut corps, otherwise he would have commanded a later mission and walked on the Moon himself. For a while he worked at the US State Department and then was the director of the world's most popular museum, the National Air and

Space Museum in Washington. There he was reunited with the Apollo 11 Command Module which he piloted around the Moon. Collins wrote an autobiography in 1974 entitled *Carrying the Fire: An Astronaut's Journeys*. It is regarded by many as the best account of what it is like to be an astronaut.

After he left the Air and Space Museum he became under-secretary of the Smithsonian Institution and then vice-president of LTV Aerospace before setting up his own consultancy. He remained married until his wife, Pat, died in 2014, aged 83.

He has admitted to having what he called an earthly ennui, though he seems to have coped with it better than most. Perhaps that was because he was able to express himself better than most of his fellow astronauts. It has been said of him that if he had ever uttered something uninteresting then no one was around to hear it.

The Apollo 11 crew, and the other astronauts, engineers, politicians, both American and Russian, who strode the space theatre of the Cold War, have left their mark on history. But as Gene Cernan once put it, those special few who went to the Moon, especially those who walked upon it, broke the familiar matrix of life, and couldn't repair it. For on the night of the climax for Apollo 11 there was, in a way, no return. No way back to their previous home. One commentator said just before the landing that we will always, somehow, be strangers to these men. There is an invisible barrier between us. They had disappeared into another life we cannot follow.

Now when I look at the Moon through my telescope I have a very different feeling from that which I experienced before man had set foot upon it. The footprints I have witnessed being made will last for tens of millions of years, perhaps longer, until bombardment by micrometeorites will churn them into the lunar dust. The descent stages will last much longer.

We have no recording of what Columbus said when he set foot on dry land after his perilous voyage of discovery, but in a hundred generations, a thousand generations, or more, someone will listen to 'that's one small step for man' crackling down the ages. Perhaps it will be heard across the galaxy.

Michael Collins wrote:

> When the history of our galaxy is written, and for all any of us know it may already have been, if Earth gets mentioned at all it won't be because its inhabitants visited their own moon. That first step, like a new-born's cry, would be automatically assumed. What would be worth recording is what kind of civilization we earthlings created and whether or not we ventured out to other parts of the galaxy.

Or perhaps when we are gone and nothing is left of mankind it will stand as to what we have done, and be what we are judged by.

For the poet Shelley the Moon came, 'from the great morning of the world when first God dawned in chaos'. Will the conquest of the Moon be a new great morning? When will we go back? After his return, Neil Armstrong wrote, 'Luna is once again isolated.' It still is.

Sources

The majority of the quotations featured in this book are from the author's personal interviews with Neil Armstrong, Charles Duke, Alan Shepard, Walter Schirra, Gordon Cooper, Walter Cunningham, Eugene Cernan, Pete Conrad, and George Low, or from NASA's Oral History projects (www.jsc.nasa.gov/history/oral_histories/oral_histories.htm).

Also included are a number of quotations that have appeared widely in space-related literature; a selected bibliography can be found on page 313.

Additional specific sources are listed below.

Prologue
5 'THEY HAD BECOME ...' *Cincinnati Magazine*, March 1972.

6 'I AM, AND EVER WILL BE ...' Speech to National Press Club, 22 February 2000.

The Spoils of War
15 'WE ARE GREATLY INTERESTED ...' *Werner Von Braun, Rocket Engineer*, Helen B. Walters (New York: Macmillan, 1964).

23 'THIS IS ABSOLUTELY INTOLERABLE ...' *Challenge to Apollo*, Asif Siddiqi (NASA publication).

29 'DEAR COMRADE STALIN ...' Baberdin writing in *Krasnaya Zvezda*, 30 April 1994.

29 'I HAD BEEN GIVEN ...' *Korolev*, Golovanov (Moscow: Nauka, 1994).

30 'WE CAN DREAM ABOUT ROCKETS ...' *The Rocket Team*, Ordway and Sharpe (MIT, 1982).

31 'IN MY OPINION IT WILL BE POSSIBLE ...' 'Soviets Planning Early Satellite,' *New York Times*, 3 August 1955.

35 'IF THE MAIN TASK DOESN'T SUFFER ...' *Memoirs*, Nikita Khrushchev (Pennsylvania State University Press, 2013).

37 'WHEN THINGS ARE GOING BADLY ...' *Target America*, S. Zaloga (Preside Press, 1993).

38 'NOBODY WILL HURRY US ...' US Dept of Commerce, National Technical Information Service, USSR Space, February 1988.

The Past and the Future
39 'OH. FRANKLY I NEVER ...' *Memoirs*, Nikita Khrushchev (Pennsylvania State University Press, 2013).

Gagarin
104 'I COULD HEAR THE SOUND ...' *Life Magazine*, 9 March 1962.

A Minor Mutiny
221 'HE CAME FLAPPING INTO MY OFFICE ...' *Last Man on the Moon*, Eugene Cernan and Don Davis (St Martin's Press, 2007).

224 'THE HATCH DESIGN DIDN'T COME INTO IT ...' *First Man*, James R. Hansen (Simon & Schuster, 2005).

Bibliography

Aldrin, Buzz, *Magnificent Desolation*. Bloomsbury, 2010

Borman, Frank, *Countdown*. Silver Arrow, 1988

Cernan, Gene, *The Last Man on the Moon*. Griffin, 2000.

Chaikin, Andrew, *A Man on the Moon*. Penguin, 1995.

Collins, Michael, *Carrying the Fire*. Farrar, Straus & Giroux, 1989.

Hansen, James R., *First Man*. Simon & Schuster, 2005.

Harford, James, *Korolev*. John Wiley & Sons, 1999.

Krantz, Gene, *Failure Is Not An Option*. Griffin, 2000.

Robinson, Jeffrey, *The End of the American Century*. Simon & Schuster, 1998.

Shepard, Alan, and Deke Slayton, *Moonshot*. Turner Publishing, 1994.

Stafford, Tom, *We Have Capture*. Smithsonian Books, 2004.

Von Braun, Ordway and Durant, *Space Travel: A History* (an update of *History of Rocketry & Space Travel*). HarperCollins, 1985.

Ward, Bob, *Von Braun*. Naval Institute Press, 2009.

Index